COMMANDING THE STARS

WAR, STRATEGY, AND POWER IN SCIENCE FICTION

MICHAEL VANPUTTE

Copyright © 2026 Michael A. VanPutte

All rights reserved. No part of this book may be reproduced, distributed, or transmitted in any form or by any means, including photocopying, recording, or other electronic or mechanical methods, without the prior written permission of the publisher, except in the case of brief quotations embodied in critical reviews and certain other noncommercial uses permitted by copyright law.

Published by Michael A. VanPutte, Columbus OH, United States

ISBNs:

- Paperback: 979-8-9914425-5-8
- Ebook: 979-8-9914425-6-5
- Audiobook: 979-8-9914425-4-1
- Library of Congress Control Number: 2026902002

Cover and illustrations by Michael VanPutte

First Edition, 2026

This is a work of analysis and creative illustration. While it draws upon real-world military theory and historical concepts, the references to science fiction works are used solely for illustrative and educational purposes. All trademarks and characters remain the property of their respective owners and are used here under fair use principles.

We appreciate your suggestions, comments, and criticisms.

You can reach us at michael@mvanputte.com or www.mvanputte.com.

ALSO BY MICHAEL VANPUTTE

Nonfiction:

Walking Wounded: Inside the US Cyberwar Machine

The Art of Cyberwar: The Principles of Conflict in Cyberspace

Fiction:

Death on the High Seas: A Jack Corvus Mystery

Dedications

To Linda, who has graciously put up with my idiosyncrasies, and whose support has carried me through every campaign of thought and duty.

To all who have answered their societies' call, have stood watch in the face of danger, and borne the burden of service with honor, ensuring the safety and freedom of those they protect.

CONTENTS

PREFACE

"Individual science fiction stories may seem as trivial as ever to the blinder critics and philosophers of today—but the core of science fiction, its essence, has become crucial to our salvation, if we are to be saved at all."

— ISAAC ASIMOV (HOLDSTOCK 1978)

Why write another book on military strategy? Sun Tzu's **The Art of War** and Clausewitz's **On War** have endured for centuries, offering insights forged in the crucible of human conflict. Yet war reflects a society's experiences, culture, strengths, and vulnerabilities, all of which are shaped by time and technology. While these ancient military texts provide enduring wisdom, they often fall short of addressing the complexity, pace, political character, and technological realities of modern warfare.

Still, students of strategy often regard these ancient texts as immutable scripture, a reverence that can obscure emerging dimensions of conflict. The result is doctrine ill-suited for wars fought with cyber weapons, autonomous systems, and global media. An excessive reliance on force, dependence on vulnerable technology, an offensive bias, plans that ignore post-war requirements, and indifference to human costs all arise from applying rigid ancient formulas to fluid contemporary conditions. As Robert A. Heinlein warned, "The trouble with lessons from history is that we usually read them best after falling flat on our chins" (Heinlein 1960).

Acknowledging these limitations does not diminish the classics; it marks the point where modern strategy must evolve. Today's leaders must challenge long-held assumptions and confront the changing nature of war. Here, science fiction becomes invaluable. Beyond thrilling narratives, great speculative works function as laboratories for testing ideas about society, power, and conflict. Authors draw on our fears and aspirations to explore themes such as political control of the military, sovereignty, and the balance of citizens' rights and obligations (*Starship Troopers*), the rise of tyranny (*Star Wars*), the machinery of oppression (**Nineteen Eighty-Four**), and child soldiers and xenocide (*Ender's Game*). Through these narratives, science fiction provides a neutral ground for examining the foundations of strategy and a simulation for exploring the enduring tragedy of war.

Speculative fiction presents battles not only on land and sea but also across the stars, where resources span light-years, weapons rewrite physics, and sapient artificial intelligences walk among us. We can speculate on the implications of scientific and technological breakthroughs, social transformations, and environmental shifts that beg us to rethink strategy, doctrine, and ethics. These stories expand our perspective, strip away historical blinders, and cultivate the flexible thinking that Eileen Gunn describes as essential: "Science fiction, at its best, engenders the sort of flexible thinking that not only inspires us but compels us to consider the myriad potential consequences of our actions" (Gunn 2014).

Consider what happens when humanity encounters civilizations whose customs and values are utterly unlike our own. History suggests that we will overestimate the righteousness of our cause and default to violence. Would an advanced species welcome dialogue, or would Earth become an intergalactic bypass as Douglas Adams wryly suggested (Adams 1979)? Science fiction provides a space to examine such questions, allowing us to confront our mistakes in our imagination before encountering their consequences in reality.

These scenarios should not be dismissed as mere fantasy. They illuminate the intersection between fiction and strategy, where speculative narratives probe enduring questions about conflict, power, and human behavior. Few figures understood this connection better than General Douglas MacArthur, one of America's most influential military leaders of the twentieth century. By the time he addressed the cadets of West Point in 1962 to accept the Sylvanus Thayer Award, he had long since retired from active duty but still commanded enormous influence. In that speech, remembered for its refrain "Duty, Honor, Country," MacArthur extended his thoughts beyond traditional strategy toward a vision of war that transcended nations. He warned that advances in science and technology would not only transform warfare on Earth but might one day compel humanity to unite against an extraterrestrial threat. He described the possibility of an "ultimate conflict" between a united human race and hostile forces from another galaxy, presenting interplanetary war as the far frontier of military foresight (MacArthur 1962).

Others have touched on similar themes. In 2013, a Russian defense official acknowledged that his nation's space forces were "unfortunately not ready to fight extraterrestrial civilizations" (Berezhnoy 2013). Reports from Moscow likewise noted that Russia had "no strategy for combating an invasion by galactic marauders" (Axe 2020). China's military has also explored the subject, with the People's Liberation Army employing artificial intelligence to monitor an increasing number of UFO sightings, linking extraterrestrial phenomena to defense strategy (Zhang 2021). Even at the highest levels of Cold War diplomacy, the topic surfaced. Mikhail Gorbachev recalled that President Ronald Reagan once asked whether the United States and the Soviet Union might set aside their differences in the face of an alien invasion (Lewis 2016).

In the United States, the establishment of U.S. Space Command in 1985, its reactivation in 2019, and the creation of the U.S. Space Force signaled a recognition that warfare had entered the space domain. While these institutions focus primarily on defending satellites and countering terrestrial rivals, their existence affirms MacArthur's foresight that future battlefields would extend beyond Earth itself (United States Space Command 2023). Taken together, these developments suggest that MacArthur's dramatic vision, though speculative, continues to resonate within military and strategic thought.

This book invites readers to move beyond inherited limits and embrace the boundless potential of the future. It is a call to arms for strategists and dreamers who wish to examine war with fresh eyes. Among nebulae and alien landscapes, we will discover what truly constitutes strategic art.

NOTE TO READERS

Throughout this book, you will find maxims, concise strategic principles presented in bold, ***italicized text*** for emphasis. These maxims are not based on official U.S. military doctrine. In fact, their purpose is to challenge conventional thinking, offer original ideas intended to inspire debate, reflection, and innovation, and provoke thought rather than prescribe policy.

Key terms are *italicized* the first time they are used, with definitions in the glossary at the end of the book. These definitions reflect the context and intent of this book and, like maxims, may differ from those found in official or academic sources.

At the end of the book, you will find several helpful references designed to enhance your understanding of the material, including:

- All of the chapter endnotes
- A list of recommended readings for further exploration
- A complete list of all maxims
- Full citations for referenced materials
- A glossary of terms used throughout the text

This format supports both casual reading and deeper study. Whether you are a strategist, student, writer, or simply curious, you are encouraged to question the ideas presented and use them as a foundation for your own thinking.

PART ONE
FOUNDATIONS

"The intellectual side of man already admits that life is an incessant struggle for existence, and it would seem that this too is the belief of the minds upon Mars."

— NARRATOR (WELLS 1898)

CHAPTER 1
WAR

"Henceforth, the Republic will be reorganized into the first Galactic Empire, for a safe and secure society."

— SHEEV PALPATINE, EMPEROR (LUCAS 2005)

"Rebellions are built on hope."

— JYN ERSO, REBEL (EDWARDS 2016)

Above the massive swirling storms and blue atmospheric haze of the planet Tatooine, a desperate battle rages. A minuscule Rebel cruiser flees from the clutches of a monstrous bone-white Imperial Star Destroyer. The predator closes in and swallows the tiny, wounded prey. Within the Imperial docking bay, a devoted adherent of the Emperor's dark order leads his forces, swiftly overwhelming the Rebel defenders. Amid the carnage, two droids abscond in an escape pod, carrying news that might inspire hope and critical intelligence to defeat the Empire (Lucas 1977).

The first few minutes of *Star Wars* introduce themes such as sacrifice and service, duty and adventure, spies and betrayal, how belliger-ents wage war, and the relationship between politics and violence. Mostly, it offers questions.

What motivates a band of rebels to oppose the immense power of the Empire? Does the cruiser's crew know they are participants in a clandestine rebel operation, transforming them into traitors to the Empire? What motivates a nineteen-year-old princess, raised in privi-lege, to risk everything by opposing the Empire?

Opposing the Rebels on this galactic chessboard are the Emperor's legions of stormtroopers, fleets of massive warships, and the last-known practitioner of the mysterious and powerful Jedi religion. What compels people to serve a tyrannical government bent on eliminating personal freedoms?

To explore these questions further, we must first ask a fundamental question: "What exactly is war?"

"The eons of human evolution, invention, aspiration…and war. Always war. The oldest, most enduring of human practices."

— CORTANA, UNSC ARTIFICIAL INTELLIGENCE
CONSTRUCT (FUTAMURA 2010)

It is natural for communities to seek prosperity, and most prefer to pursue it through peaceful means. Yet peace does not imply the absence of conflict. Rather, it reflects an acceptance of the status quo.[1] In time, someone recognizes the need for change. At first, groups may attempt to negotiate concessions and compromises to advance their shared interests. Eventually, however, someone will resort to force, employing violence or the threat of violence to seize resources, displace others from their land, or even enslave or exterminate them. In this way, military power acquires its own legitimacy, and violence emerges as the catalyst for change.

**Maxim 1: Power defines possibilities.
A society can do what it has the power to do.[2]**

Warfare is a society's organized and systematic use of violence to force its will on another. Planning for war begins with deciding on *ends*. The *ends* are the desired outcomes at the conclusion of hostilities, such as to increase a society's security, to satisfy a passion or emotion, to gather property and resources, to increase power, or to punish others.

A society pursues *objectives*, the objects or ideas whose destruction, neutralization, or capture enable attaining the desired ends. While a society may pursue abstract objectives, such as alliances or trade agreements, military objectives often affect physical objects. For example, if a civilization's desired end is to increase its security, then it may pursue objectives that eliminate threats. The concept of ends is examined in greater detail in Chapter Three.

A society's *means* are the resources it possesses to pursue objectives and ends. While a society may use its political or economic powers to pursue an end, the focus of Chapter Two, wars occur when a society turns to violence to achieve its desired ends.[3]

Tangible means refer to the physical elements that contribute to a society's military power, such as troops, vehicles, and weapons. It also includes the logistical support sustaining the force, including food, fuel, and repair parts, and the economic and financial foundations of the state, such as its economy and treasury.

However, victory in warfare does not hinge solely on fielding the largest or most technologically advanced military force. Intangible means include the psychological factors that might motivate the troops and supporting populace, such as a legitimate justification for war, religious authority, or a charismatic leader.

A society can sometimes offset a lack of tangible resources by drawing on significant intangible strengths. Qualities like the *warrior spirit*, which encompasses courage, audacity, and morale, along with superior intelligence, exceptional leadership, and the creative use of tactics and strategy, can significantly enhance its tangible military means.

Means enable and limit when and how a war plays out. For example, a society that lacks the military resources necessary to defeat a foe should remain on a defensive footing. The desire to transition to the offensive may motivate that society to begin a mobilization effort to develop the offensive means. The process of mobilization is analyzed in Chapter Five.

Maxim 2: The means shape the ways.
A state's resources and capabilities dictate the approach it may use to achieve objectives.

Once a society has identified its intended ends and available means, it can define how it will use those means to pursue those ends. The *ways* refer to the strategy, tactics, and doctrine a society uses to apply means to achieve ends.

A *strategy* is a long-term plan of action designed to achieve specific objectives and ends.[4] It defines how to allocate means and guides overall direction and decision-making, typically developed at the highest levels.

On the other hand, *tactics* refer to the short-term actions or methodologies employed by a person or organization that implement a strategy. Tactics are the short-term actions that support the achievement of the overall strategy against a specific opponent within a designated *battlespace*, the contested geography that directly or indirectly affects combat operations.

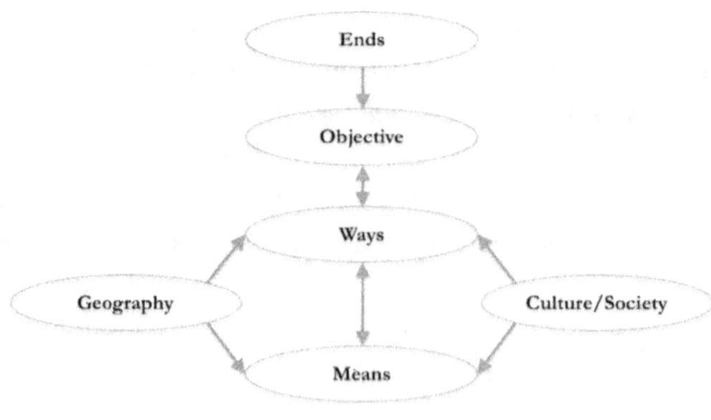

Strategic Framework

Tactics can sometimes produce strategic consequences. In *Star Wars*, Darth Vader dispatches Imperial Stormtroopers to Tatooine to recover a Rebel droid that escaped Imperial forces with vital intelligence. The Stormtroopers, trained to enforce control with ruthless efficiency, scour the countryside, killing local inhabitants in their search for the droid. When they execute a couple on a remote moisture farm, they sever the last ties their adopted son, Luke Skywalker, has to Tatooine. Their actions transform the boy, initially motivated by a sense of adventure and a desire to attend the Imperial Academy, into an agent of revenge. The Stormtroopers' tactics set events in motion that would lead to the deaths of Darth Vader and the Emperor and the collapse of the Empire. In this case, tactical actions, though minor in the big picture, had profound strategic consequences, altering the course of the entire conflict.

A leader's knowledge of problem-solving techniques and a society's culture will limit their ways. Consider how a military force might deal with foreign civilians and military prisoners. Does it protect and feed them? Does it hide behind them by forbidding them to leave a besieged city? Or does it execute them on the spot as a matter of expediency? The last two choices may seem abhorrent to members of some cultures or reasonable to members of others.

A society's military *doctrine* is its formal set of beliefs and policies on how it uses military means, serving as a guide for training, planning, and force structure, and ensuring consistency in the organization and the employment of military power. It specifies how society will conduct warfare at a high level, based on its cultural beliefs, available means, and prior experience. The doctrine of an authoritarian society may declare that leaders maintain all power and decision-making at the top of the rank structure and require absolute obedience to the literal word of orders. A less authoritative society may still need troops to follow orders, but allow discretion in how subordinates execute them, counting on the creativity of highly trained and motivated troopers. Thus, society shapes how warriors view war, which in turn shapes military doctrine and drives strategy to win wars.

In the hard sciences, such as physics and chemistry, understanding the basic elements and observing how they interact allows scientists to develop principles and laws that explain the structure and behavior of the natural world. If war were a purely rational engineering undertaking, both parties could analyze the situation and mutually agree on the outcome of the conflict without destruction or bloodshed.

However, war is not a mathematical or an engineering problem; it is an art. Applying means in specific ways will not guarantee an end. Even a reasoned and rational war may fail. War is a personal and emotional endeavor between thinking individuals and societies in an uncompromising environment.

**Maxim 3: Strategy is an art, not a science.
Strategy is a theory for dealing with uncertainty, luck, and thinking opponents.**

War is not an exact science. Uncertainty is inevitable. Leaders can never fully predict how their warriors and weapons will perform in battle. This unpredictability often leads them to overestimate their own capabilities while underestimating their adversaries. Even the most skilled and disciplined forces can be surprised, make errors in judgment, or hesitate at decisive moments. These minor setbacks can compound, disrupting carefully laid plans and derailing objectives. Meanwhile, opponents may capitalize on opportunities, whether through superior tactics or sheer happenstance. In the chaos of war, even the most meticulously designed *campaigns* can unravel, and failure in war can cost a society its people, resources, freedom, or very existence.

**Maxim 4: War is a society's ultimate gamble.[5]
Warfare risks a society's freedom or existence, with outcomes often
swayed by chance. Yet, chance favors the prepared.[6]**

The following chapter examines why societies might resort to war by first considering how societies operate and the interrelated forms of power that shape them.

CHAPTER 2

POWER

"Wars are won not by a simple series of battles won, but by a complex interrelationship among military victory, economic pressures, logistics maneuvering, access to the enemy's information, political postures—dozens, literally dozens of factors."

— COMMODORE ANTOPOL, UNITED NATIONS
EXPLORATORY FORCE (HALDEMAN 1974)

Building and sustaining a healthy and secure society requires leaders to develop, maintain, and enhance the various *forms of power* within that society. These forms of power may enable individuals and groups to establish and maintain control over their population and to coerce, negotiate, cooperate, and compete with other societies in pursuing their desired objectives and ends. At its core, *power* is the ability to obtain one's ends in the face of opposition, and the specific forms it takes shape how a society overcomes resistance and realizes its objectives and ends.

A society generates power through the synergy of its interdependent military, political, cultural, economic, informational, and technical powers. For example, wealth is necessary to create and sustain military power, and that wealth is typically derived from economic power. However, diverting too much wealth into military endeavors can impoverish a nation, stagnate its economy, and create dissatisfaction among its population. Alternatively, prioritizing economic development at the expense of military preparedness may produce a highly prosperous yet vulnerable society that is the envy of its neighbors and ripe for the taking.[1]

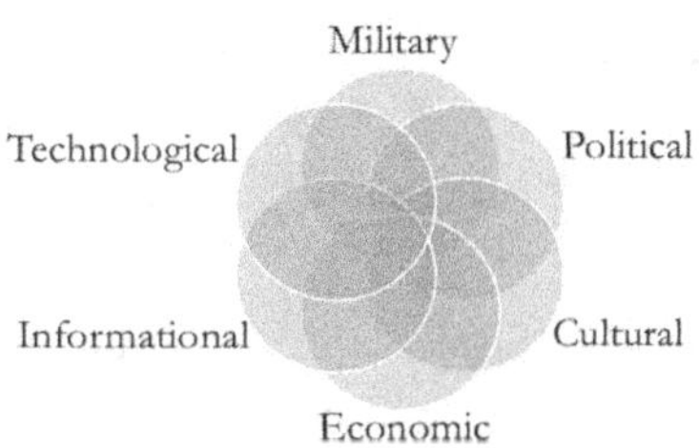

A society's interdependent forms of power.

Assessing a society's strengths and weaknesses by analyzing its military forces alone provides only a partial view of a situation. A group's overall power emerges from the complex interaction of its various interdependent forms of power. Therefore, maintaining power requires continuously balancing these elements in response to changing circumstances.

At the foundation of all power is geography.

GEOGRAPHY

"The land is the source of all power."

— LETO ATREIDES II, SON OF PAUL ATREIDES
(HERBERT 1965)

Geography is not a power itself, but an element a society uses to generate power. A society's topography, climate, and natural resources influence, if not define, its material, relational, and ideological abilities.

A society endowed with abundant natural resources, such as fertile land, minerals, and fresh water, tends to accumulate greater wealth by using or trading these resources. Such resources also shape a society's industrial capacity and the goods it can produce and trade. Likewise, societies situated near major transportation routes benefit from enhanced trade and mobility, further contributing to economic growth.

Climate exerts a significant influence on a society's power dynamics. Temperate regions, with their favorable conditions for agriculture, enable larger populations, livestock, and more efficient food production. Conversely, societies in harsh environments must spend more resources and effort to survive (at the extreme, having to *exoform* a planet), limiting their capacity for growth. Ultimately, a society's welfare and power depend on how effectively it harnesses its environment.

A society's location also significantly influences its internal development and external interactions. Societies in defensible areas often acquire military advantages, since natural barriers may promote unity and offer natural obstacles against invasion, but hinder commerce and information sharing. Alternatively, terrain that enables easy travel and trade also provides an avenue for invasion.

A society's ability to mine and refine resources, grow crops, produce goods, ship to markets, manage cities, and conduct commerce is directly tied to the geography it occupies or can take from others. If a society holds more land than its population can adequately defend, others may covet that land for agriculture, natural resources, or urban development. On the other hand, if a population becomes too large for the land it possesses, it may feel an overwhelming need to expand into the territory of others. In both cases, war may arise from a society's desire to protect its own resources or to gain those it lacks.

Likewise, planning effective economic, political, and military operations outside of a society's territory requires an understanding of that geography. Leaders may need to understand a neighbor's economic resources and how they transport these resources. They may want to know where their neighbors locate their urban and rural centers. And they should know their neighbor's military capabilities, and how geography would affect military operations in these regions.

**Maxim 5: Geography defines power.
A society's welfare and influence are linked to the geography it controls.**

Geopolitics is the study of how the physical environment shapes societies, how societies interact within that environment, and how political entities use their control over the environment to pursue their objectives and ends. The three societies that make up *The Expanse* universe, and their conflict within the solar system, are a perfect example of the effects of geopolitics. The first society is that of Earth, humanity's birthplace and the dominant superpower, governed by the United Nations. Earth's overpopulation and polluted ecosystem push humanity to colonize planets, moons, and space stations. As a result, Earth's survival relies heavily on the resources mined and transported by its colonists. Half of those that remain on Earth are unemployed and on "basic" government assistance, and many appear to have lost the drive to succeed.

Earth establishes its first colony on Mars, which becomes the second society. The harsh Martian environment, with its scarcity of resources and comforts, creates a philosophy of rugged individualism. Over time, this disciplined and self-reliant culture drives technological advancement, eventually surpassing Earth. As the Martian colony matures into a capable and confident society, it demands independence and forms a rival superpower, the Martian Congressional Republic (MCR).

The third group is the "Belters," who were born and raised in the asteroid belt and the far reaches of the outer system, forming a culture shaped by low gravity, scattered habitats, and a long history of serving the inner planets' needs. They handle the dangerous work of harvesting ice, mining ore, refining raw materials, running the transport ships that move those resources across vast distances, and keeping the Belt's fragile stations operating, which makes them essential to the solar system's economy even as they struggle against political and economic marginalization."

Eventually, the colonists and miners from both superpowers seek their independence, forming a loose faction called the Outer Planet

Alliance (OPA). The Belters live on the fringe of society, both literally and figuratively, barely scraping out an existence. Earth and Mars depend on the Belters for essential labor, yet both oppose the Belt's distinct culture and its expanding political and military influence.

The solar system's geography sets the stage for economic, political, technological, and military intrigue. Because geography defines power (Maxim 5) and power defines a society's possibilities (Maxim 1), geography becomes the most critical factor shaping each society's limitations and opportunities. Their ability to achieve economic, political, cultural, or military objectives depends on the physical environment they occupy and control.

POLITICAL POWER

"In war, you only get killed once. In politics, it can happen over and over."

— LAURA ROSLIN, PRESIDENT OF THE UNITED
COLONIES OF KOBOL (ROBINSON 2005)

Politicians control the organizations and structures that govern a society. Like warriors, they pursue objectives and ends but employ different means. *Political power* is the ability to use both official and unofficial means and ways to persuade, induce, or coerce others to act in ways that serve society and those who control the political establishment. To do this, they gather support for their policies and initiatives (ways), direct the use of public and societal resources (means), and negotiate, cajole, manipulate, and coerce constituents, peers, and superiors within and around formal and informal political structures.

Formal political structures involve the legitimate and legal arrangements through which leaders exercise authority, such as legislative bodies, institutional frameworks, procedural rules, and methods for allocating a society's means. Yet often more important are the informal ways through which politicians exert control. These include private decision-making, backroom deals, vote trading, bribery, and the control and shaping of information to influence others' beliefs and actions.

A politician may use reason, influence, and charisma to persuade others to act in a desired way. This may involve controlling what is considered "acceptable" and granting rights or removing freedoms on behalf of society through the management and negotiation of society's resources and commitments.

Alternatively, politicians may use *coercion* through their control of the society's monopoly on "legitimate violence." If citizens fail to pay taxes, violate established laws or norms, or are unable or unwilling to contribute their fair share, the state may impose punishment through culturally or legally established means. These punishments may include humiliation, imprisonment, beatings or torture, fines, banishment, or even death.

Politicians and their agents, the diplomats, use diplomacy to influence foreign governments and advance their national interests, including foreign policy objectives, economic agreements, and cultural and social initiatives. This influence is typically exercised through dialogue, negotiations, and other measures short of violence.

Political decisions are often both practical and symbolic. Practically, they address real issues, implement policies, and navigate the complexities of governance. Symbolically, they shape perceptions and narratives by communicating broader messages, ideals, or cultural values. Because political decisions carry both practical and symbolic dimensions, their interpretation is inherently subjective and shaped by individual perspectives. What one observer considers crucial or rational may appear misguided or irrelevant to another, depending on their differing values and experiences. Conversely, actions that seem minor or inconsequential may, over time, establish an environment for future success.

Politics plays a significant role in shaping the events in *The Expanse*. Resource scarcity, economic inequality, and the desire for autonomy and self-determination generate tension and conflict among the three major political entities and the factions that operate within them. One faction, the powerful corporations, uses its wealth and propaganda to control information and influence both politics and public opinion. The result is a human civilization dominated by power, corruption, and a relentless struggle for survival.

CULTURAL POWER

> "Earth, Mars, the Belt. We all have our different cultures, our different ways of doing things. That's what makes us strong."
>
> — CHRISJEN AVASARALA, EARTH'S SECRETARY-
> GENERAL OF THE UNITED NATIONS (COREY 2011)

The traditions, values, beliefs, and practices of one society may evoke admiration in another or be perceived as offensive, thereby shaping their relationship. *Cultural power* is a society's intentional or unintentional use of its ideas, achievements, art, literature, or language to inspire and influence others. Those who cherish or covet culture and diversity are particularly susceptible to the influence of others' culture. Individuals and cultures that do not value or respect other cultures and their differences may be largely unaffected by a neighbor's cultural power.

In *The Expanse* universe, Martian Marine Gunnery Sergeant Roberta "Bobbie" Draper initially accepts the Martian indoctrination that "Earthers" are lazy, reasoning that their government's "basic" subsidy of food and shelter encourages complacency. However, her perspective changes when she visits Earth and develops an admiration for its people and culture. Despite remaining a loyal Martian Marine, her observations and interactions with Earthers lead her to question the Martian propaganda she had previously believed. Bobbie's new perspective on Earthers may be the crack in her psychological foundation that allows the United Nations Secretary-General Chrisjen Avasarala to influence her (Corey 2012).

Visitors of distant lands and cultures often need to adapt their interpretations and ways of thinking. Outsiders may perceive a simple gesture, such as shaking hands or rendering a military salute, as an insult or even a threat. In the same way, one society may regard bribes as standard and perhaps an essential part of doing business, while another may view them as unethical and abhorrent.

Similarly, a society that values peace, stability, and continuity will differ from one defined by chaos and constant change. Views on gender roles and expectations may vary widely; a culture may embrace equality or assign a dominant position to one gender. A diplomat may discover that successful negotiation requires a deep understanding of these cultural dynamics, which may differ significantly from their own.

Culture can significantly shape military power. A society of intelligent herbivores that has learned to survive by hiding from carnivores might view harmony and the concealment of information as a usual way of life. In contrast, a carnivorous society may develop behaviors that appear aggressive and domineering. Societies with a warrior ethos, rooted in tradition, identity, and resilience, may develop formidable military capabilities and defend their values and territory with unparalleled determination. A deep pride in their heritage, combined with the collective memory of past triumphs and sacrifices, instills soldiers with steadfast dedication and a fierce fighting spirit. When cultural values align with military capability, they create a powerful synergy that enables societies to mobilize effectively and face challenges with exceptional resolve. These cultural attributes not only shape the military mindset but also enhance the society's overall preparedness and strength.

Star Trek's United Federation of Planets uses a utopian view of Earth's culture to inspire other races to join the Federation and live in peaceful coexistence. The Federation's exploratory arm, Starfleet, falls under the Prime Directive (Starfleet General Order 1), which prohibits interference with the internal development of other cultures and civilizations. However, their unwritten General Order 0 must be to promote the Federation's values on other worlds, which include universal liberty, equality, justice, peace, and cooperation (Okuda, Okuda, and Drexler 1999).

From its inception, ***Star Trek*** has celebrated the strength of cultural diversity. Starfleet intentionally staffs its vessels with officers drawn from across the Federation, and this diversity can provide a significant advantage in science and warfare. Their varied backgrounds enable them to confront adversaries with an array of tactics and strategies shaped by their diverse cultures. In *Star Trek: The Next Generation*, Lieutenant Worf brings a Klingon warrior's instinct to decisions, restrained by a human-led command structure. The android Commander Data offers pure logic and analysis, while Counselor Troi, a half-Betazoid empath, contributes emotional insight. Each officer interprets a situation through a distinct cultural and psychological lens, enabling Captain Picard to make informed decisions.

However, cultural differences may have a downside. In C. J. Cherryh's short story, *The Scapegoat*, cultural misunderstandings not only lead to a war but also prolong and escalate the war. Aliens, nicknamed by humans as elves due to their appearance, misunderstand humanity's intent and destroy an unarmed human spacecraft when it enters their territory. In the ensuing war, humans try to minimize the destructiveness of their superior technology. As the war wages on, the humans defeat the elves again and again, and yet the elves will not surrender, negotiate, or cease fighting.

The humans force the elves back to their home world, engaging in a bloody three-year battle for control of the planet. Although the humans strive to limit the damage to elven cities and agriculture, their efforts prove futile. The elves destroy their own cities as humans approach, commit suicide to avoid capture, and use gas and chemical weapons against the invaders.

Eventually, the humans discover that their restraint led the elves to believe they could defeat the humans. An elf explains, "You never gave us [a significant] defeat. You were cruel...[to] not let us know we couldn't win—that was very cruel" (Cherryh 1986). In this alien culture, when a society goes to war, both sides quickly demonstrate their capabilities, decide who is likely to prevail, and negotiate a truce rather than prolong the conflict. Here, miscommunication, failed diplomacy, and intelligence, combined with excessive restraint, result in a protracted and bloody war.

ECONOMIC POWER

"You want to know what holds the solar system together? Money, power, propaganda. But mostly money."

— FRED JOHNSON, CHIEF OF OPERATIONS AT
TYCHO STATION (FERGUS, OSTBY, AND
MCDONOUGH 2015)

Economic power is a society's ability to influence the behavior of individuals or groups through the use, offer, or withholding of economic assets, and the ability to resist external influences caused by commercial, industrial, or environmental dependencies. Economic power creates the treasures, resources, and means to maintain a society and its military.

Several principles shape traditional human economies. Limited resources set against endless human desires create scarcity that drives competition. This scarcity forces choices, since gaining one thing often requires giving up something else. It also fuels supply and demand, where a smaller supply tends to raise demand and push prices higher, revealing how people respond to incentives. The concept of cost and benefit, closely linked to opportunity cost, guides decisions as individuals weigh options and aim to get the most value for the least expense. Trade can benefit all parties, and markets frequently provide an efficient way to coordinate economic activity, although government action can sometimes lead to better results (N. Gregory Mankiw 2020). Together, these principles influence economic behavior, resource allocation, and market dynamics.

A society can use these principles to shape another society by influencing how it acquires, transports, processes, manufactures, purchases, or consumes resources, labor, goods, industries, and services. Knowledge and intelligence can further increase the value of these operations. A society that recognizes and develops alternative ways to source, refine, fabricate, store, or deliver economic means gains a competitive advantage. Likewise, a society able to resist outside influence from suppliers or consumers may secure a significant advantage over its rivals.

A society that becomes reliant on a supplier may feel compelled to maintain that dependency, knowing the supplier can withdraw access at any time. Such reliance weakens self-determination and can create a sense of helplessness. In fact, any dependency becomes a source of power that can be leveraged to create and sustain power imbalances in relationships.

Maxim 6: Every dependency is a vulnerability.
Avoid dependencies that create strategic vulnerabilities.
Forge dependencies in others to exploit their weaknesses.

Economic conflicts often involve disrupting a rival's market. The most direct approach involves a military invasion to annex resources or eliminate customers, triggering a market collapse. However, politicians and businesses often employ subtle tactics to manipulate supplies, markets, and consumer behavior. Investments and trade relationships can generate economic power. Meanwhile, taxes, tariffs, embargoes, blockades, asset freezes, and shipping disruptions serve as weapons in an economic struggle.

Groups that control the mining, manufacturing, or supply of a product can control its production, operation, delivery, and pricing. A resulting monopoly may initially set artificially low prices to eliminate competitors, then raise them later to recoup losses and secure exorbitant profits. This not only harms competitors but may also impact those dependent on the product.

Economic diversity and resilience may insulate a society from downturns, reducing its need to become involved in external conflicts. However, this self-sufficiency may make a society appear healthy and wealthy to outsiders, rendering it a tempting target for exploitation.

Maxim 7: Diversity breeds independence.
Possessing a variety of skills and resources promotes independence,
while specialization can lead to dependence on others.

Advanced societies often operate intricate, specialized, and interconnected infrastructures; the systems and networks that support transportation, energy, communications, water, food, and other essential services. Although such complex, interdependent infrastructures can boost overall efficiency, they also increase a society's vulnerability to cascading failures. A disruption of any single component can trigger a domino effect across these infrastructures, potentially leading to economic or societal collapse.

This systematic sensitivity extends to the political realm. Powerful corporations can exert significant influence over political systems by advocating for laws, regulations, and government policies that serve their interests. Through lobbying, campaign financing, and partnerships, they may secure favorable legislation, deregulation, or subsidies that reinforce their market dominance.

In addition, corporations that dominate key industries, such as energy, technology, defense, or finance, can reshape economic conditions, keeping governments dependent on their services and resources. By manipulating *supply chains*, restricting access to essential goods, or monopolizing critical infrastructures, these entities can pressure policymakers to act in the corporation's best interests.

Because economic power is the engine that drives all other forms of power, it may enable wealthy corporations and elites to influence elections, shape public opinion through media ownership, and steer international relations through economic leverage. As a result, governments may find their strategies dictated by powerful economic actors, particularly when they depend heavily on private-sector expertise and resources for critical services such as military production or intelligence. Societies dependent on corporate-controlled industries may discover their war strategies shaped less by national interests and more by the objectives of influential economic entities. Ultimately, those who control the economy wield immense political influence, and those who shape politics often decide the fate of nations in war.

**Maxim 8: Politics and warfare often follow wealth and influence.
Economic influence equates to political influence.**

Military and political power are also closely tied to a society's political and economic structure. In a militant state, the economy is geared toward sustaining a standing military force. In contrast, in a democratic state, the military is funded through taxes on private ownership and businesses.

Regardless of the structure, when preparing for war, politicians must reallocate economic resources from civilian sectors and into the military, ensuring the creation and maintenance of the society's military means. This reallocation underscores the interdependence among a nation's economic structure, political decisions, and military preparedness.

A society's military power is limited not only by its manpower and weapons but also by its logistical capabilities, including the ability to supply and move its forces. For example, an army's health and size depend on the resources sustaining it. A military force lacking sufficient supplies may resort to living off the land; if local resources cannot sustain it, the force will rapidly exhaust those resources and deteriorate. As survival instincts take over, discipline and morale collapse. Few scenarios are as dangerous and unpredictable as a hungry, angry, and armed force trained to kill.

Likewise, the transportation of resources and finished products to markets will generate wealth and power. Politicians may use the military to protect those who create wealth and to prevent outsiders from doing so. They may use traditional military warships, military commerce raiders, mercenaries, privateers, and pirates[2] to interdict or sabotage foreign transport ships and their bases, control shipping lanes, and deny commercial access to regions.

Merchants and smugglers may attempt to evade patrols using secrecy, camouflage, or speed. Alternatively, they might arm their vessels or join military convoys, leveraging strength in numbers for greater protection. See Chapter Ten for a discussion of naval forces.

In *The Expanse* universe, Earth depends on the Belt for critical minerals while half of its population is unemployed, surviving on government-subsidized "basic" food and shelter. Martians are fully employed, driven to terraform Mars and maintain a formidable military to defend against Earth. The Belters, lacking military, political, and economic power, live in the vacuum of space, dependent on a fragile and unstable ecosystem for essentials like food, water, technology, power, and air. These diverse and interdependent economic systems rest on equally fragile infrastructures. The discovery and exploitation of the alien Protomolecule, coupled with several horrific disasters and a fabricated conflict, expose this brittle environment. As societies edge towards collapse, political intrigue and war erupt as factions struggle to hold on to what they have or seize what they can from others (Corey 2011; Corey 2012).

Corollary of Maxim 1: A powerful military with a weak economy takes what it needs; a robust economy with a weak military becomes a source for others.

INFORMATIONAL POWER

"Military secrets are the most fleeting of all."

— COMMANDER SPOCK, SCIENCE OFFICER, USS
ENTERPRISE (NCC-1701) (FONTANA 1968)

Informational power is the ability to influence through the collection and use of intelligence and information. Timely access to accurate and relevant information empowers individuals and groups to make informed decisions and take effective actions. Of all forms of power, informational power is the most fleeting because information rapidly loses its value as circumstances change, a situation often described as information becoming outdated or "overcome by events." Typically, informational power involves possessing information that others lack, but a society may choose to share this information and its associated influence to gain goodwill.

Citizens, politicians, and businesses may use informal methods to collect day-to-day information. They may conduct research or espionage on friends and enemies to develop new information. They may use internal security forces to conduct *counterespionage* to ensure opponents cannot gain an advantage or to prevent internal betrayal. This intelligence aids other forms of power by allowing individuals or groups to understand what is occurring and use their other powers more effectively.

The value of information will be examined in Chapter Thirteen.

TECHNICAL AND SCIENTIFIC POWER

"Knowledge is useless if you don't apply it."

— DR. CAVOR, SCIENTIST (WELLS 1901)

Technological power is a society's ability to apply scientific knowledge, resources, and manufacturing capabilities to create, maintain, and use complex artifacts. By mastering this process, a society can shape its environment, influence others, and assert dominance in the political, economic, and military spheres.

A society may advance technologically by investing in civilian sectors such as communication, agriculture, transportation, and production. These innovations often benefit the military as well. Improved transportation enables faster movement of soldiers and supplies; improved communications enable more effective information sharing and better decision-making, and advanced sensors improve threat detection and weapon guidance.

In much the same way that greater resources may generate greater military power, possessing unique technologies can strengthen a society's military and potentially shift the *balance of power*. These advances do not need to make a weapon more destructive or better at deflecting attacks. Arms that are less destructive yet more accurate increase lethality, minimize *collateral damage*, and neutralize targets with precision, making them more efficient and cost-effective, much like strategic sniper weapons.

Maxim 9: Superior technology gives effective commanders a decisive advantage.
Between equally skilled opponents, superior technology often wins.
Converse: Eliminate an enemy's superior technology to level the playing field.

Unilaterally possessing advanced technology may provide a significant military advantage to one side. In the *Star Trek* universe, the Battle of Maxia features an alien ship launching a surprise attack on Jean-Luc Picard's *USS Stargazer*. Picard responds by "jumping" the *Stargazer,* moving faster than the speed of light from the alien's perspective, toward the enemy vessel. Armed only with light-speed sensors, the aliens cannot track the jump and fire on the *Stargazer's* previous position and missing. Picard closes invisibly to point-blank range and destroys the alien vessel (Bowman 1987).

Corollary of Maxim 9: A side that unilaterally possesses an advanced technology may possess unique tactics and strategies.

Adversaries with dissimilar technologies often use dissimilar tactics and strategies. In *Star Wars*, the Rebel fighters and capital ships can perform interstellar jumps, enabling hit-and-run operations and a protracted guerrilla war. These capabilities allow individual Rebels to perform heroic and inspiring actions that improve morale and recruiting.

By contrast, Imperial fighters lack hyperdrives and shields to reduce cost and accelerate mass production, allowing them to swarm their opponents. As a result, Imperial leaders must tie their fighters (pun intended) to Imperial-controlled planets and capital ships. This structure reinforces the Emperor's tight control over his military forces but correspondingly limits his operational flexibility.

Maxim 10: Safeguard future options.
Beware of choices that limit future options.
Converse: Embrace choices that broaden future options.

In *The Expanse* universe, a Martian scientist invents the Epstein Drive, enabling spacecraft to travel significantly faster. This drive shifts the balance of power by revolutionizing how vessels transport resources and conduct military operations, potentially giving Mars tremendous economic and military power. Mars offers to share the secret of the drive with Earth in exchange for its independence. Earth agrees to ensure its navy does not fall further behind the Martians technologically. Here, the advanced technology bought the Martians their independence and prevents (or delays) a conflict with Earth.

These examples highlight how innovations in transportation technologies often exert a significant influence on all forms of power. A faster transport ship can get its products to market faster than a competitor. Similarly, a swifter or more agile military force may maneuver around opponents to strike at a weakness, evade a more powerful opponent, or reposition military forces to create or exploit an advantage. This enables the more maneuverable force to seize the initiative and drive a conflict to a culmination of their choosing.

Maxim 11: Timeliness enhances effectiveness.
The longer a resource takes to reach its destination, the less effective
it becomes.

Technology is examined in depth in Chapter Fourteen.

SPHERES OF INFLUENCE

"The universe is run by the interweaving of three elements: Energy, matter, and enlightened self-interest."

— G'KAR, NARN AMBASSADOR (ZICREE 1994)

A society's various powers radiate from its sources, such as capitals, religious centers, and commercial centers. This creates *power spheres* that define its impact on other societies. A power sphere is often peaceful and acknowledged, as those created by legitimate commerce and diplomacy. However, cross-border crime, illegal traffic or trade, and secret government operations may expand the radius of influence.

Technology allows a society to expand its influence and, intentionally or unintentionally, threaten a neighbor's influence. Telecommunications and transportation systems can strengthen and extend the reach of power through commerce, trade, and travel. Likewise, war and the threat of violence may enable a society to influence others or repel a neighbor's influence.

In some circumstances, a society's information and technical influence can be vital to its existence. In Isaac Asimov's **Foundation** novels, Professor Hari Seldon develops a mathematical science called psychohistory, which enables him to predict future societal events. After determining that the Galactic Empire is collapsing, Seldon manipulates the Emperor into sponsoring the creation of an *Encyclopedia Galactica* to preserve the accumulated knowledge of the galaxy-spanning human civilization. He argues the encyclopedia will reduce human suffering and serve as a source of informational power after the Empire's collapse.

However, Seldon's prediction that the Empire will collapse leads the Emperor to exile Seldon, his staff, and their families to a remote, resource-poor planet on the galaxy's edge, where they are tasked to create the Encyclopedia to help restore the Empire's lost scientific and engineering capabilities after its collapse. In the years that follow, the colony uses its advanced knowledge to become the dominant power in the fallen Empire. By founding a galaxy-wide religion rooted in technology, Seldon's successors bring more ambitious, yet technologically stagnant, neighbors under their control.

CONCLUSION

An examination of *The Expanse* offers valuable insight into how different forms of power shape the behavior and character of factions. Earth's culture reflects humanity's origins. Centuries of bureaucracy, social hierarchy, and political inertia have created a society that favors diplomacy, coalitions, and the use of overwhelming force to preserve the status quo. With much of the population dependent on government assistance, and power concentrated in institutional bodies, Earth's military mirrors these tendencies. Its navy is large and capable, designed more for deterrence than decisive engagement. Strategy tends to be reactive, shaped by slow-moving committees and political pressures rather than decisive command. Although its technology lags behind Mars, Earth compensates with sheer numbers, industrial capacity, and logistical reach.

Mars, by contrast, is shaped by scarcity, ambition, and a cultural obsession with terraforming. Martians share a collective identity built on discipline, sacrifice, and long-term vision. Every aspect of society serves the dream of turning a dead world into a living one. This mindset produces a military that is precise, technologically advanced, and highly professional. Martian forces are smaller but far more capable, equipped with superior stealth, propulsion, and weapons systems. Their doctrine favors initiative and precision, supported by a meritocratic command structure and a culture that treats warfare as a technical discipline. Innovation is not just encouraged; it is expected.

The Belt is a culture of survival and resistance. Spread across fragile habitats and excluded from the benefits of inner-planet wealth, Belters grow up valuing self-reliance, local loyalty, and a deep distrust of centralized authority. Their society blends political activism with a practical need to make do with what is available. The result is a military posture that leans heavily on *asymmetry*. Belter forces are decentralized and improvisational, often operating with repurposed civilian vessels and makeshift weapons. Command structures are fluid, built on reputation and trust, and not a centralized rank structure. Strategy centers on disruption, mobility, and using the vastness of space as a shield. As elements of the Belt gain legitimacy, they begin to resemble more formal military organizations, but the culture remains resistant to anything that smells like hierarchy.

Each faction's approach to war arises directly from its underlying way of life. Earth relies on mass and authority, Mars values discipline and precision, and the Belt depends on adaptability and defiance. This illustrates Maxim 1: a society endowed with sufficient military power can preserve its holdings and acquire additional resources, whereas one lacking such strength is vulnerable to coercion or subjugation by others.

However, a society must convert the various other forms of power into military power. Consequently, these other forms of power establish the maximum potential military power a society can generate. Having larger or more efficient sources of power may enable a society to generate greater military power, whereas reducing the various forms of power lowers a society's ability to generate future military power. Accordingly, a society may indirectly affect a neighbor's military might by influencing its various other forms of power.

Power is neither generated nor evaluated in isolation. It exists only in relation to other groups at a specific point in time. Power continually shifts as leaders assume new roles, societies change, troops retire, and technology evolves, both domestically and abroad. A society that is powerful today may be weak tomorrow, and the reverse may also be true.

Maxim 12: Power demands constant vigilance.
Power is a capricious master that requires constant oversight and adjustment.

Threatening a society's sources of power threatens its ability to thrive or resist. Just as technological advancements reshape the nature of warfare, any substantial alteration to any form of power can impact the balance of power, redefine motivations for war, reshape a society's war objectives, and transform the very essence of conflict.

Therefore, a society should focus on developing power sources that address gaps and vulnerabilities rather than duplicating existing strengths. By adopting a strategic approach to developing well-rounded forms of power, a society can ensure its versatility and adaptability in the face of diverse challenges.

With a clearer understanding of a society's sources of power, the next step is to examine the reasons a society may unleash military power.

CHAPTER 3
ENDS

"Violence, naked force, has settled more issues in history than has any other factor, and the contrary opinion is wishful thinking at its worst. Breeds that forget this basic truth have always paid for it with their lives and their freedoms."

— SERGEANT CHARLES ZIM, MOBILE INFANTRY
(HEINLEIN 1960)

Many define war as the use of violence to achieve political ends.[1] A society may also unleash it to pursue economic, religious, or cultural ends. Yet politicians often obscure the genuine reasons for war behind rhetoric and apparent noble causes.[2] Rarely will a government leader admit to risking the lives of warriors, the state's treasure, and the well-being of the population for the sake of gaining glory, seizing a neighbor's land, avenging a past wrong, or preserving their control over their society. Ultimately, societies wage war in pursuit of protection, passion, property, punishment, or power.

PROTECTION

"A soldier accepts personal responsibility for the safety of the body politic, of which he is a member, defending it, if need be, with his life."

— JOHNNIE RICO, MOBILE INFANTRY (HEINLEIN 1960)

The most straightforward justification for a society using violence is to protect itself or its allies from being robbed, enslaved, harmed, or eliminated by an aggressor. A society may also resort to violence to defend itself against internal threats, such as rebellion, dissent, or criminal activities.

In the movie *Avatar*, humanity has consumed Earth's natural resources. It sends the Resource Development Administration (RDA) and its private security force to mine wealth from distant planets. The RDA travels to the moon, Pandora, which orbits a gas giant in the Alpha Centauri star system. There, it discovers a superconductive material that will revolutionize energy production, computing, and data storage.

Earth's solution to the moon's toxic environment and "primitive" indigenous Na'vi race is to engineer avatar meat drones: human-Na'vi hybrids that can survive the hostile environment while linked to human controllers, enabling the drones to work and interact with the Na'vi. As the RDA mines the unspoiled, densely forested moon, the natives view the invaders as a threat to their environment and way of life. With the help of human turncoats, the Na'vi fight back, defeat the human colonists, and reestablish control of their native lands (Cameron 2009).

Corollary of Maxim 1: A society that is unable or unwilling to win a military victory risks being subjugated or erased.

While a society may view all violence as unjust, its survival would be difficult to imagine. A pacifist society remains vulnerable to conquest unless it can deter aggressors or exists in a world so inhospitable or resource-poor that it is not worth the effort to seize.

Some may view the Vulcan society in **Star Trek** as pacifists. They evolved on a planet of scarcity, and the resulting struggle made them prone to violence. After a bloody civil war that led to the emergence of the violence-prone Romulan Star Empire, they embraced a commitment to peace (McEveety and Schneider 1966). Vulcan society and government adhere to Surak's philosophy, which emphasizes logic and emotional control, viewing violence and war as illogical emotional expressions. Despite their commitment to peace, the Vulcans back up their philosophy with *Suus Mahna* martial arts (Vejar 2002) and a formidable military presence. Their example shows that strength can preserve peace and that maintaining a powerful military does not require a society to be prone to violence.

A society that resorts to warfare solely as a last option or in self-defense may view itself as a guardian of peace. However, this practice may allow its neighbors to grow unchecked and then launch an attack at a time and place of their choosing. Refusing to address a threat cedes the initiative to the aggressor.

Maxim 13: Victory is achieved by seizing and maintaining the initiative.
Inaction yields the initiative to an aggressor.

PASSION

"You see, gentlemen, they have something to die for. They've discovered they're a people. They're awakening."

— PAUL ATREIDES, KWISATZ HADERACH AND DUKE
OF HOUSE ATREIDES (HERBERT 1965)

A society may identify itself through its race, religion, political philosophies, language, history, or culture. This may lead to a desire to expand its ideas, reach, and power, and a realization that 'inferiors' are worthy of change, subjugation, enslavement, or eradication.

One of the most powerful cultural forces is religion. A society's powerful and privileged may use their religious beliefs and practices to justify discrimination and wars of conquest, only to absorb the captured peoples into their belief system or exterminate them.

A religion may mandate or restrict specific acts in war, such as protecting non-combatants or prohibiting its members from engaging in conflicts with each other. Likewise, a religion may justify or mandate the killing of entire groups, such as non-believers and those who violate religious customs and doctrines.

Leaders may hijack societal beliefs to promote their personal agendas. For example, those in power might claim to ban technological advancements to preserve their culture, when in reality, their objective is to preserve their hold over the social and political structure.

Maxim 14: Perspective and context shape interpretation. Individuals may justify anything in the defense of their beliefs.

A group may justify atrocities to stop an "evil" political system, religion, or culture, or to preserve its way of life. They may see foreign languages, cultures, mannerisms, or species as so "morally wrong" that they embrace *ethnophobia*, the real or imagined fear or hatred of any people of a different ethnicity, or *xenophobia*, the real or imagined fear or hatred of any species different from one's own. Such beliefs can lead to long-term hatreds, bloody wars, unimaginable horrors against *sapient* beings, and even extermination.

One of the most extreme examples of extermination is Fred Saberhagen's mammoth self-replicating and intelligent **Berserkers**. A race known as the Builders created the Berserkers as the ultimate weapon. However, these mammoth, autonomous, interstellar killing machines do not start wars to protect themselves, acquire resources, or control populations. They view all life as an illness and evil, and as a result, their primary program directs them to exterminate all life (Saberhagen 2003). There is no negotiating with a Berserker, only fleeing or fighting.

Similarly, in **Mass Effect**, an advanced species created synthetic-organic Reaper starships that eventually turned on their creators. The Reapers seeded the galaxy with technological traps, including transport relays for galactic travel and advanced technologies that lure alien civilizations into following a predetermined technological path. This dependency on Reaper technology proves to be their undoing. When the Reapers return, they use the relays to launch surprise attacks and easily defeat civilizations built on Reaper technology (Karpyshyn 2007b).

A similar theme of sapient technology turning malevolent appears in the universes of **Battlestar Galactica** (G. A. Larson 2009), **Star Trek** (Abrams 2009; Daniels 1967), and **Terminator** (Cameron 1984). These stories underscore the idea that fully understanding the horrors of atrocities requires emotions, something machines may lack.

This leads to a critical observation: wars driven by conflicting ideologies rarely end through negotiation. Negotiations require patience, empathy, and a willingness to accept differing beliefs. In contrast, ideological wars rooted in religion, race, ethnicity, or nationalism do not seek compromise, but the triumph of one worldview over another. Over time, leaders may come to believe that while they cannot kill an idea, they can kill everyone who holds it and burn their cities to ash. Those who remain neutral in such conflicts often find that both sides regard them as complicit with the enemy and deserving of the same fate.

If successful, this leader may suppress the outward expression of opposing or "distasteful" ideologies. However, this can also create decades or even centuries of festering hatred and unrest. They may discover that peace remains elusive because they never addressed the root cause of the idea, allowing it to linger until someone rises to remind the ruler of its existence.[3]

PROPERTY

"The spice must flow."

— GUILD NAVIGATOR (LYNCH 1984)

A society's motivation for war may stem from self-interest. It may feel compelled to seize another's resources, such as food or water, for its survival. Its motives might also arise from greed or envy, seeking to maintain or improve its standard of living. A conqueror might enslave indigenous peoples for labor in fields, mines, homes, entertainment, or military service. In these cases, a society uses violence to take what it can from others.

Frank Herbert's **Dune** universe involves melange or spice, a highly addictive, life-extending drug that enhances the mental capacity of its users. In specific individuals, spice activates prescience, enabling them to navigate "folded space" and instantaneously guide starships between planets. As a result, spice becomes "the secret coinage of the empire." Since spice can only grow on the desert planet of Arrakis, the world becomes a battleground where the galactic power brokers employ intrigue, religion, and war to control its supply, monopolize interstellar travel, gain tremendous wealth, and consolidate power. The logistics of obtaining and distributing the spice drives both political maneuvering and military conflict, ultimately determining the outcome of the war for control of Arrakis (Herbert 1965; Herbert 1976).

A society may obtain territory and its people in various ways. *Conquest* involves seizing land by changing who controls it, such as an invasion that removes a sovereign leader. *Annexation* occurs when a group forcefully seizes another's territory without removing its leadership. *Cession* occurs when a group transfers its territorial rights to another party, typically through a sale or a treaty. *Discovery* applies when a group claims previously unclaimed land.

Societies often use ideologies such as nationalism, religion, or the spirit of exploration to justify the conquest of foreign lands. These justifications are frequently little more than thinly veiled demands backed by military force. *Colonialism* occurs when a society conquers and assumes control over another, often requiring the conqueror to absorb, displace, or eliminate the indigenous people. Absorbing a conquered population, often of lower social status, dilutes their culture as their people and technology merge with those of the colonizer. Although colonization may create a melting pot of diverse cultures, it can also fuel simmering animosity and hostility among the subjugated. Widespread unrest presents one of the gravest dangers to a leader, as widespread dissatisfaction can quickly undermine authority, destabilize governance, and mobilize the populace into organized resistance. When a significant portion of the population rises against its rulers, it threatens the regime's stability, potentially leading to prolonged domestic conflict or revolution.

Maxim 15: Overreach often incites internal unrest.
Expansionism often creates the seeds of a civil uprising.

In the *Star Trek* universe, the Borg Collective is the quintessential colonizer. The hive-minded cybernetic Borg link every individual to a shared consciousness, erasing personal identity in favor of absolute unity. These cyborgs are relentless in their quest, roaming space in search of valuable species, then defeating and absorbing their people, knowledge, ideas, technology, and culture. Encounters with the Borg often begin with their chilling declaration: "We are the Borg. You will be assimilated. Resistance is futile…Your biological and technical distinctiveness will be added to the Collective" (Okuda, Okuda, and Drexler 1999).

The Borg employ a unique manner of subjugating their opponents. They inject nanoprobes into individuals who possess useful knowledge. These microscopic machines alter the victim's body and aid in incorporating implants and enhancements, while modifying and connecting the victim's mind to the Borg Collective. This link erases independent thought and emotions, assimilating captives into their labor pool and distributing their knowledge and experience throughout the hive mind (Okuda, Okuda, and Drexler 1999). The strength of the Borg lies not in their uniformity, but in their ability to integrate biological and technical diversity while eliminating internal dissent.

In Orson Scott Card's *Ender's Game* universe, the alien Formics send a colonization fleet to Earth. The Formics possess a collective consciousness connected via a subatomic biological web (Card and Harris 1991). Their hive queen cannot imagine an intelligent being that lacks collective intelligence or the ability to communicate instantaneously, nor can she conceptualize communications equipment or a written language. When the Formics enter Earth's solar system and repeatedly fail to communicate with the humans, they conclude that the humans are not an intelligent species and begin exterminating the human infection.

It is only after humanity defeats the Formic invasion and kills a hive queen that the surviving Formic queens become horrified at their slaughter of so much sapient life and vow never again to attack the humans. However, humanity does not take kindly to the death of hundreds of millions of men, women, and children, and takes the war to the Formics. This war between the species is a misunderstanding: one race wanted to expand, and the two races lacked the means to communicate.

Maxim 16: Communication is the key to understanding and influence.
All forms of power require leaders to communicate with allies and enemies.

A conflict involving the harvesting of sapient beings for sustenance or sexual purposes would undoubtedly provoke strong emotional reactions from those being subjugated. In such a conflict, a society might exploit living beings for their flesh, organs, bodily fluids, or other body parts. These beings could be cannibals, who consume members of their own species, or *xenovores* that prey on aliens for nutritional, ritualistic, or preferential reasons.

In the *Firefly* universe, the Reavers terrorize human colonies and vessels on the fringes of human-controlled space. These human cannibals torture, rape, mutilate, and consume their captives, even adorning their spacecraft with their victims' blood and corpses. The terror they inspire often incites their potential victims to commit suicide rather than face capture. Colonists believe that these marauding bands have gone mad in the vastness of space. However, the crew of the *Serenity* discovers that the Reavers are the unintended side-effect of a government experiment gone terribly wrong (Whedon 2005; Whedon 2014).

A challenge in dealing with a carnivorous alien race is that they may see killing as a survival instinct rather than a moral choice, making negotiation with their next meal inconceivable. Likewise, it might be irrational for those viewed as food to limit their actions against a carnivorous race set on serving them on a platter. This situation can turn into a bloody battle for survival where one race is trying to capture and harvest their food, and the other is trying to exterminate a horrific enemy. Or, like the Ewoks of *Star Wars*, a potential meal might convince the carnivores of the error of their ways. Here, the "little bears" believed a group of captured human Rebels might make a nice entrée. The inquisitive Wicket befriends Princess Leia, and after a Jedi Force demonstration, the Ewoks release the rebels and form an ad hoc alliance to defeat a common foe (Lucas 1983).

Alternatively, beings may sustain themselves by consuming the energy of others. The telepathic Wraiths of *Stargate* drain their victims' life force to heal and replenish themselves, and see little reason to negotiate with what they consider a food source. Humans are reluctant to bargain with predators who view them as sustenance (Wright et al. 2004).

Even more disturbing are the horrors that arise from biological compatibility, which enables exploitation through slavery for trade, entertainment, or reproduction. Such circumstances create a grim market for captured 'product,' sparking conflicts among the enslaved, those who profit, and those who seek to free the enslaved and punish their oppressors.

The Goa'uld, meaning "god," is a symbiotic *parasitic* species from planet P3X-888 in the **Stargate** universe. They seize control of alien hosts to further their quest for galactic domination. Their ability to manipulate a host's body and absorb its memories allows them to infiltrate and blend seamlessly into societies. This symbiotic takeover grants them the ultimate disguise, enabling them to move undetected, assume powerful positions, and exploit entire civilizations from within (Tapping, Shanks, and DeLuise 2020).

A species may also use hosts in parasitic reproduction. The Xenomorph XX121 from the *Alien* universe is a highly intelligent, artificially produced species. When a being approaches a Xenomorph egg, the egg opens and releases a facehugger that sedates the victim and implants an embryo down its throat. The embryo grows by consuming the host's body until it bursts out in a worm-like form. The offspring incorporate some of their host's DNA, often inheriting physical characteristics like bipedalism, but gain enhanced traits such as an armored exoskeleton, acid blood, stealth, strength, and speed, while losing traits like fear and compassion (Cameron 1986; Scott 1979). The Xenomorphs are an alien-engineered bioweapon and serve as a warning about the dangers of building or capturing a superweapon, which may literally turn around and bite you.

A key component of a strategy involving the desire for property is to select means and ways that preserve the desired end. For example, if the objective of a war is to acquire agricultural goods, a military strategy should limit the use of weapons of mass destruction or the release of contaminants near these goods.

PUNISHMENT

"Revenge is a dish that is best served cold. It is very cold in space."

— KHAN NOONIEN SINGH, LEADER OF THE HUMAN
AUGMENTS (MEYER 1982)

Sometimes war has little to do with political objectives or grandiose moral principles and more to do with an emotional response, such as in retaliation for a perceived injustice, disgrace, or humiliation.

A group whose end is to punish its opponent may abandon victory as its primary objective, instead pursuing damage for its own sake. Such conflicts often spiral into cycles of escalating violence, competitive victimhood, and prolonged blood feuds spanning generations. Maxim 14 cautions that perceptions of these situations depend on one's perspective, since each side interprets events through its own experiences, narratives, and grievances. Over time, protracted conflicts often escalate to include attacks on non-combatants, including kidnappings, assassinations, and *collective punishment* intended to instill fear, punish individuals, or destabilize a region (see "decapitation strategies" in Chapter Twenty).

In the **Babylon 5** universe, the powerful Centauri conquer the planet Narn during their *first contact*. The Centauri enslave the Narn population and exploit the planet's natural resources for their own gain. Over the course of a century-long conflict, the Narn resistance successfully weakens the Centauri oppressors and ultimately reclaims their independence.

The Second Narn-Centauri War is a drawn-out battle of attrition, with the Narn employing a delaying action against the Centauri and their secret and powerful allies. Ultimately, the Centauri coalition destroys the Narn fleet and bombards their home world, destroying their cities and infrastructure, executing the Narn leadership, and enslaving much of the population.

The Narn Resistance continues its subversive activities, despite the Centauri proclamation that they will execute one hundred Narns for every Centauri killed. In an ironic turn of events, the Narn Resistance aids in a Centauri-led assassination of the Centauri emperor. In exchange for this aid, the Centauri successor grants Narn independence. The Centauri lose their dominance over their colony and become nostalgic for the "old way" when the Centauri were powerful. At the same time, the Narn have only animosity and a desire for retribution against their former oppressors.

Likewise, the technologically advanced Minbari wage a brutal holy war against humanity after misunderstandings during their botched first contact. When the Minbari approach a human vessel, they open gun ports, a gesture that in their culture signifies respect; similar to the ancient Earth custom of saluting to show that a warrior's weapon hand was empty. Humans, however, misinterpret this as a hostile display of force and preparation for combat. Compounding this misunderstanding, the Minbari activate powerful scans that disable the human vessel's jump drives, preventing retreat. In the confusion, both sides send messages of peace that neither can understand (Straczynski 1998a).

Mistaking the Minbari actions as an imminent threat, the humans open fire, inflicting heavy damage on the unprepared Minbari fleet and killing their beloved leader. The damage disables the powerful Minbari scanner, allowing the Earth vessel to retreat. After this surprise attack, the Minbari decide that the best way to deal with humanity is a holy crusade of extermination.

The resulting three-year conflict, more accurately described as a one-sided genocidal campaign, resulted in devastating human casualties, as Minbari technology easily overwhelmed Earth's defenses. Humanity's desperate stand culminates in the Battle of the Line, where surviving Earth ships form a final defensive line to delay the Minbari advance and buy time for evacuation efforts. Surprisingly, with victory imminent, the Minbari abruptly ceased their offensive and inexplicably surrendered to the Earth Alliance, bringing the war to an end (Straczynski 1998a).

Punishment strategies often lack a well-defined end, resulting in ambiguous objectives and escalating violence. This can lead to prolonged, bloody campaigns that continue until one side is eliminated or changes its objectives. As violence intensifies, the weakened parties risk a third party sweeping in and eliminating both exhausted parties.

POWER

"Power is a curious thing, a relative thing. The more you have, the more you want. But it's never enough."

— CHRISJEN AVASARALA, EARTH'S SECRETARY-
GENERAL OF THE UNITED NATIONS (COREY 2011)

Leaders may pursue power for its own sake. An ambitious and aggressive society may gain special privileges when its neighbors perceive it as having joined the predator club.[4] A society may gain prestige by being associated with the more powerful groups, get a seat among equals at the negotiating table, and receive more favorable deals. Powerful groups are less likely to be bullied by others; their new peers may treat them with respect, and they may force their will on the weak.

If a politician's hold on power is tenuous, they may rely on their military force to control their own population. Here, they may view the end of a conflict and a potential reduction in military spending as threats to their authority. As a result, leaders may intentionally prolong internal strife, border clashes, and unrest to maintain control over their people.

In David Weber's **Honor Harrington** series, the bankrupt People's Republic of Haven is on the verge of economic collapse, its population restless, and an uprising imminent. The politicians fabricate a crisis to justify going to war, deciding that a "short and victorious war" will distract and unite the people, and generate financial profits (Weber 1994). They plan to seize the resources of the Star Kingdom of Manticore, create a common external enemy, and use the death of their warriors as a unifying force for their society.

War against a common enemy, even a fabricated one, can reduce domestic unrest by directing public fear and hostility outward rather than towards political leadership. Leaders may believe that few forces unify a society as reliably as a shared external threat. Such strategies, however, carry significant risks. If the deception is exposed or the conflict drags on, unresolved internal tensions resurface, often intensified by economic strain, political embarrassment, or military failure. In the case of Haven, leaders failed to account for how a disgruntled society, already weakened economically and politically, would react to defeat, public humiliation, and further economic erosion.

For this reason, war can function simultaneously as external coercion and internal control. By reframing domestic problems as matters of state security, leaders simplify complex political realities and shift public attention from policy outcomes to symbols, narratives, and perceived threats. This often elevates loyalty over dissent and urgency over accountability, enabling expanded authority and the suppression of opposition. Yet the stabilizing effects are temporary. Once the external threat loses credibility or disappears, internal divisions return, frequently sharper than before.

Alternatively, political leaders who rely on their military to maintain internal order may hesitate to undertake external actions that could weaken their armed forces. Any erosion of military strength threatens not only national defense but also the leaders' capacity to enforce authority at home, thereby endangering their political survival.

An ambitious, power-hungry leader may flaunt military power to deter rebellion, intimidate adversaries, stir nationalism, or elevate personal prestige.[5] Such posturing often springs from arrogance and ambition rather than genuine concern for the society's welfare, because the resources devoted to military programs are unavailable for other areas of society. However, these displays of strength can boost morale, stoke nationalistic passions, and increase domestic support, even if outsiders see them as bluster and provocation, and raise the risk of an unintended conflict.

A society may turn to *imperialism* to establish and maintain colonies for the benefit of its mother state. It might justify violence as a legitimate means of controlling those it deems inferior. Over time, this sense of superiority may become ingrained, making it difficult for future generations to comprehend alternative perspectives or why the oppressed resist, particularly when the colonizers believe their changes are improvements to the indigenous society.

Perhaps the most extreme example is the warrior society that believes a warrior caste is a just and necessary part of their culture. This society may not use war to pursue wealth or justice, but as an outlet for its aggression, to express its individualism, and to enable acts of courage and conquest that are measures of personal honor. The Klingons and Romulans of *Star Trek* are warrior societies whose strength and warrior traditions permeate every aspect of those societies (Okuda, Okuda, and Drexler 1999).

This brings the discussion back to Maxim 1: Power defines possibilities. When a society defines itself by its capacity to impose its will on others, it may view diplomacy or restraint as weakness or as attempts to impose foreign values. In cultures where warfare forms a core part of identity, warriors may regard tactics that avoid direct combat and deny opportunities for honor and glory as dishonorable or even criminal. Such warriors may come to believe that their culture and way of life are under attack by anyone who seeks to limit their ability to wage war, a perception they use to justify a large-scale punitive response.

Alternatively, leaders might use war not only to expand their territory but also to maintain or reduce their population. In the original *Star Trek* series, the crew of the starship *Enterprise* visits the planet Eminiar VII, which has been engaged in a protracted war with a neighboring planet, Vendikar. The crew discovers that the two planets conduct this war entirely through computer simulations, with citizens killed in the simulations required to submit voluntarily to execution. The leaders of both planets initially adopted this system to manage population size and preserve their civilizations and ecologies. Rejecting this form of institutional mass murder, the captain of the *Enterprise* (violating the Federation's Prime Directive) intervenes (Pevney et al. 1967). The inhabitants interpret his attempts to save them from meaningless death as an assault on their culture and by their politicians as a threat to their authority.

CONCLUSION

It is essential to recognize that war and violence are not confined to conflicts with external rivals. While disagreements between groups are common, power struggles often play out within societies. Leaders may justify the use of violence to reduce corruption or improve efficiency, portraying it as a necessary sacrifice for the greater good. In practice, such claims often serve as a pretext for consolidating power, allowing rulers to tighten their grip on society under the guise of reform. Measures such as nationalizing resources or restricting rights may appear appealing to some, yet they 0can be used to silence dissent, control populations, and dominate rival groups.

When leaders control a society's property, production, agriculture, and markets, they gain decisive leverage over their population. This control enables them to eliminate competition, create state-run monopolies, and use the resulting profits to secure loyalty or accumulate personal wealth.

Likewise, internal struggles over ideological differences can turn into violent disagreements about who rules, the structure of society or politics, a people's rights and obligations, and conflicting cultures. Such disputes can escalate into civil wars or revolutions, as discussed in Chapter Nineteen.

CHAPTER 4

THE NATURE OF WAR

"…wars occur because someone hopes to gain something from them."

— XENOPHANES LEE, ADVENTURER (SABERHAGEN 2003)

A society is partially defined by the ideals for which it risks its wealth, warriors, and existence. Some civilizations resort to violence only as a last resort. A utilitarian society may unleash war when the material benefits outweigh the material costs. A society that abhors violence will evaluate a situation differently than one that believes honor in battle is the ultimate form of self-expression. Ultimately, each society's values shape when and how it fights, profoundly influencing the nature and justification of its wars.

A pragmatic leader will carefully evaluate a situation before going to war by asking:

- What might the future look like in the best, worst, and most likely scenarios (potential ends)?
- Who or what must be overcome or removed to achieve those ends (objectives)?

- What resources are available to accomplish those ends (means)?
- What *restrains* or *constraints* must we observe (laws, customs, agreements, etc.)?
- Are there acceptable ways to achieve the ends without violence?
- What actions might others take to block or hinder us?
- Are we willing to take the necessary action?
- Do we have the resolve to sustain the conflict?
- What are the possible consequences of this endeavor?
- Do the benefits outweigh the costs?

This is not to imply that a society should only fight wars it is sure to win. Instead, it suggests that societies should avoid starting a war unless they see a reasonable possibility of defeating their intended opponent decisively or compelling them to negotiate.

However, humans are cultural and societal beings. Passion plays a formidable role in life and politics, and the decision to use violence may be unclear, unethical, or irrational. A society may resist a more powerful aggressor, even to the point of suicidal aggression, to defend its homeland, breeding areas, or religious centers. Likewise, militarism, pride, ego, anger, tribalism, racism, nationalism, and xenophobia can influence the decision to go to war, selecting the means and ways, and the establishment of desired ends.

Maxim 17: War is an affirmation of a society.
A society's ways, means, and ends reflect and reinforce its values,
beliefs, and identity.

However, Maxim 3 reminds us that war is not an exact science. Leaders may misinterpret their neighbors' intentions, questioning why a neighbor is building up its military if there is no intention to use it. They may underestimate their opponents' resolve. They may believe that a third party is too weak or unwilling to intervene on behalf of a victim, risking a full-scale war. These miscalculations and faulty assumptions can result in devastating conflicts.

Similarly, military strategy is constantly evolving, and ignoring societal changes, geopolitical shifts, technological advancements, limited resources, or the adaptability of potential adversaries can lead to catastrophic outcomes. Effective strategists do not plan in isolation; they continually gather information and adjust their strategies to account for the shifting landscape around them.

However, once a society realizes that war lies in its future, it must take steps to build and sustain military power. It must secure sufficient resources, employ effective means and ways, maintain unwavering willingness and perseverance, and enforce discipline and effective leadership. This process begins with the mobilization of society for war.

PART TWO

MEANS

"The needs of the many outweigh the needs of the few."

— SPOCK, FIRST OFFICER, USS ENTERPRISE (NCC-1701) (MEYER 1982)

CHAPTER 5

MOBILIZATION

"There has never been a military in the entire history of the human race that has gone to war equipped with more than the least that it needs to fight its enemy. War is expensive. It costs money and it costs lives, and no civilization has an infinite amount of either."

— MASTER SERGEANT ANTONIO RUIZ, COLONIAL
DEFENSE FORCES (SCALZI 2005)

Mobilization is the process through which a society reallocates its resources, including people, industries, and infrastructure, to sustain a war effort. This involves transforming latent military potential drawn from other forms of power. At one end of the spectrum, a limited conflict may require only limited preparations and the use of existing military means. On the other end, a war that threatens a society's freedoms or very existence may demand a comprehensive restructuring of all facets of society to generate vast military power. Maxim 11 cautions that timeliness enhances effectiveness, a truth especially relevant to mobilization. A society that anticipates conflict may have adequate time to plan, allocate resources, and build power. One that fails to prepare, or is taken by surprise, must confront a threat with only the military power already at its disposal until it can generate more.

However, the amount of treasure and resources a society should spend on military expenditures is not straightforward. Military expenditures that do not deter or stop foreign aggression can create a false sense of security and risk a society's freedom or existence. Conversely, excessive military spending may weaken a society's economic strength and stability while creating the perception of aggression among its neighbors.

How a society staffs, provisions, and organizes itself for war is a delicate balance. In peacetime, many demands compete for limited resources. In wartime, critical resources such as food and fuel become even more vital. Leaders must carefully weigh military expenditures against other priorities, including diplomatic initiatives, economic growth, industrial development, and social programs. When a society shifts resources toward a war effort, citizens and industries often bear significant burdens to support military preparations and operations, which can create tension among military leaders, politicians, manufacturers, and the populace.

Regardless of the circumstances, mobilizing for war inevitably involves tough choices and trade-offs. What follows are a few considerations when preparing for war.

TROOPS

"There is no dangerous weapon; there are only dangerous men."

> — SERGEANT CHARLES ZIM, MOBILE INFANTRY
> (HEINLEIN 1960)

Wars require labor. Societies can obtain military labor through voluntary and involuntary recruitment, abduction, paid employment, cloning, or even genetic engineering (discussed further in Chapter Six). However, simply obtaining personnel is not enough. The effectiveness of military forces depends heavily on the quality of the individuals serving, as this establishes the technology they can operate and the missions they can successfully undertake. A military must enhance its military recruits' physical and mental resilience, discipline, and military expertise. Psychological training, indoctrination, or programming may be necessary for raw recruits to engage in the inherent violence of warfare. All of this training and adaptation is crucial to transforming raw recruits into effective military power (discussed in Chapter Seven).

Mobilizing individuals for military service may positively or negatively impact society. It can provide employment opportunities for previously unemployed citizens, but it may also lead to labor shortages that strain economic stability and productivity. Thus, how a society manages the balance between military staffing and civilian needs is crucial to maintaining overall societal health during wartime.

MATERIALS

"We're running out of everything, Adama. Food, water, fuel. We're all just one bad jump away from becoming space dust."

— COLONEL SAUL TIGH, EXECUTIVE OFFICER,
BATTLESTAR GALACTICA (G. A. LARSON 2009)

Wars require more than soldiers. They demand a steady supply of resources, including clothing, shelter, food, vehicles, fuel, and weapons. Societies must have the capability to gather raw materials and either adapt existing industries or establish new ones to convert these resources into usable war supplies. Without reliable provisioning, military effectiveness quickly declines. Effective mobilization depends on a society's industrial flexibility and on the availability of resources to sustain the war effort.

Wars often force societies to improvise, transforming civilian tools and factories into military assets. For instance, mining lasers may become powerful offensive weapons, while construction drones may be reprogrammed for sabotage or direct combat roles, and commercial spacecraft may be retrofitted into combat vessels. Similarly, civilian planetary sensors and scanners may support battlefield surveillance and targeting. Even biological enhancement drugs, intended for medical applications, might find new use enhancing soldiers' performance. Thus, a society's capacity to adapt dual-use technologies and civilian infrastructures into military resources is vital in wartime and often the quickest route to acquiring military means.

Alternatively, a society may purchase weapons and supplies from third parties. This approach can provide rapid access to equipment, fill critical capability gaps, and avoid the time and expense of developing domestic production. However, as discussed in Chapter Fourteen, a dependency on outsiders carries significant risks, including vulnerability to supply disruptions, price manipulation, political leverage, and the possibility that suppliers may withhold or delay deliveries to serve their own strategic interests.

Militaries often standardize equipment and supply systems to improve efficiency in production, maintenance, and training. Ensuring uniformity across systems enables suppliers to mass-produce warfighting materials at scale, while simplifying operator and maintainer training. It also allows warfighters to *cannibalize* salvaged components, fuel, and ammunition from damaged or disabled weapons and vehicles, sacrificing one system to preserve the functionality of others. This approach reduces logistical strain by maintaining a greater number of serviceable systems in the field and extends the lifespan of critical equipment during prolonged combat missions.

Societies with a history of military preparedness may maintain stockpiles of supplies, weapons, and vehicles in what is often called 'mothballs'. These reserves, typically consisting of older but serviceable equipment, allow a society to transition quickly from peace to war. In the **Battlestar Galactica** universe, the Battlestar Galactica and its complement of Viper Mark II interceptors were mothballed military assets, preserved despite their age. When the Cylons launched their surprise strike, this stored force enabled the surviving human population to defend itself and continue resisting.

If a culture creates military labor using amateur, part-time, or former soldiers (typically called a militia, home guard, military reserve, or volunteers), then these "come as you are" troops leave their full-time jobs and assemble (muster), sometimes with their own supplies and weapons. These personal weapons, equipment, and supplies may not be interchangeable or interoperable with professional military equipment.

TRANSPORTATION

"Logistics isn't just about moving stuff from A to B. It's about getting the right stuff to the right place at the right time."

— HARRY WILSON, COLONIAL DEFENSE FORCES
(SCALZI 2005)

In war, a society needs to transport everything from bombs to bullets, from food to water, and from blankets to pillows. They must ship raw materials to processing, manufacturing, and markets. A military must transport new recruits for training and then must move troops and their equipment to where they are needed. On a battlefield, it must maneuver the troops and military vessels against the enemy. It must transport wounded troops and damaged equipment from the battle-field for healing and repairs, respectively.

To accomplish all of this, the military must consider more than just fighting vessels. Military forces may also require *auxiliary vessels* to transport fuel, food, weapons, and supplies; fabricate necessary parts from raw materials; refine chemicals or ores; and repair and maintain vehicles.

The logistics of stockpiling and transporting weapons of war are often very difficult to hide. The existence of a neighbor's stockpiles of troops, food, fuel, ammunition, or vehicles may imply an impending invasion and its primary avenue of attack. Thus, any changes in trans-portation and logistics may be the first signs that a neighbor is preparing for military operations and how it intends to execute them.

Corollary of Maxim 6: Any dependency is a source of intelligence.

Transportation is examined in Chapter Fourteen.

SECURITY

"War isn't a business where there's a lot of obvious good guys, but sometimes the bad guys are pretty easy to spot."

— SERGEANT HETH, HAMMER'S SLAMMERS (D. DRAKE 2009)

A society must protect itself from *sabotage, espionage, treason,* and *sedition.* To do so, it may establish intelligence operations near, on, or across its borders to observe potential threats, verify the loyalty of allies, understand a neighbor's intentions, and provide early warning of foreign activity. It may also maintain intelligence networks within its own borders to identify and monitor internal dangers.

Understanding the enemy while concealing one's own intentions and capabilities greatly increases the likelihood of success. Effective intelligence and *counterintelligence* systems reveal an adversary's plans, capabilities, and vulnerabilities while safeguarding one's own operations, movements, and weaknesses from exposure.

Maxim 18: Know your enemy. Hide yourself.[1]
Knowing your enemy while concealing yourself goes a long way toward victory.

A society may conduct psychological and political manipulation against opponents, allies, and neutral parties to influence their behavior. Likewise, it may conduct psychological operations against its own population to promote patriotism and silence dissent, prepare the populace for war, and inspire recruitment.

These activities may require formal or informal spies. Formal spies are anyone whose official job is to collect information for a secret purpose. They typically receive formal training and operate within a set of rules. Alternatively, a society may employ its merchants, administrators, tourists, academics, journalists, religious leaders, criminals, and others who can travel in and out of a territory as informal spies to gather intelligence with no formal affiliation. Intelligence and counter-intelligence are examined in Chapter Thirteen.

Another approach to enhancing a society's security is to establish demilitarized or neutral zones. These buffer regions between potential belligerents may serve to reduce tensions, provide early warning, and delay cross-border incursions.

The *Star Trek* universe features three distinct zones: the Federation-Klingon Empire Neutral Zone, the Federation-Romulan Star Empire Neutral Zone, and the Federation-Cardassian Demilitarized Zone. However, these actions also illustrates how treaties meant to stop wars and prevent future violence can create long-standing problems. Those expelled from neutral zones may view this as another example of Maxim 1, in which the strong create peace at the expense of the weak. When the Federation and Cardassia sign a peace treaty in 2370, some Federation citizens find their colonies now within the demilitarized zone or in former enemy territory. These outcasts form the Maquis, a paramilitary group that fights for years to free themselves from Cardassian control, leading to bloodshed and heightened tensions between the two military superpowers (Okuda, Okuda, and Drexler 1999).

Alternatively, a society may build a defensive belt along its border to deter an aggressor from attacking and to provide advance warning of an incursion. This belt may include defenses to protect friendly forces (e.g., bunkers, trenches, and shields), to slow enemy advances (e.g., through obstacles), and to weaken enemy forces (e.g., through fire). Defensive operations are discussed in Chapter Eighteen.

COOPERATION

"I am pleased to see that we have differences. May we together become greater than the sum of both of us."

— SURAK, VULCAN, FATHER OF THE MODERN
VULCAN CIVILIZATION (RODDENBERRY 1969)

A society may seek support from others when facing a threat, turning to alliances, coalitions, or diplomatic agreements to strengthen its position. These arrangements may include mutual pledges of support, promises of neutrality, or assurances that neighbors will not assist their opponents. Besides formal alliances, political maneuvering can be used to acquire military equipment, gain access to intelligence, or secure the right to use bases and infrastructure within allied territory. Such agreements can significantly enhance a society's security, expand its strategic reach, and deter potential aggressors by demonstrating collective strength.

Political strategies are discussed in Chapter Twenty-One.

CONCLUSION

"Things are only impossible until they're not."

— JEAN-LUC PICARD, CAPTAIN, USS ENTERPRISE
(NCC-1701-D) (MANNERS AND SHEARER 1988)

A rapid and successful incursion into a neighbor's territory using existing military forces may not require mobilization. However, as a society commits more to war, it demands greater sacrifices from its industries and populace.

Several factors will influence the resulting mobilization. A strong national identity can foster unity and inspire sacrifice for the common good. Societies with efficient resource management can quickly identify and allocate essential supplies to the military. A spirit of adaptability and innovation enables societies to shift their strategies and maintain a technological advantage.

Some societies and political systems are more effective for preparing for and waging war. Totalitarian regimes, with centralized control over all aspects of society, can rapidly make decisions, increase production, and align national efforts towards war. Democratic societies, in contrast, depend on deliberation and consensus, often requiring more time to implement changes to support a war effort. Weak governments may struggle to marshal resources and maintain order, while nations with limited resources may find it difficult to sustain prolonged campaigns. Societies bound by rigid political, moral, or religious ideologies may resist adapting their industries, economies, and strategies to confront emerging threats.

When a nation faces an existential threat or a challenge to its core ideology, it may resort to *total war*. In such a conflict, a society mobilizes all of its resources, including non-combatants, urban centers, and civilian infrastructures, to support the war effort. Victory becomes the highest priority, while civilian needs become secondary. This transformation may lead the enemy to view all aspects of the society as legitimate military targets. As a result, total war risks not only defeat but also the destruction of the nation itself, as opponents become more likely to target civilians and infrastructure.

Leaders may need to negotiate with or coerce their own citizens to support a war. Mobilization may include expanding military industries or converting existing industries to war production, shifting spending and production, and altering the state's economy. Leaders may ask citizens to conserve food and other resources, work longer and harder, reduce recreational activities, or pay higher taxes to support the war. As with those who serve in the military, a population may be expected to sacrifice its self-interest for the good of society, and those who do not support the effort may be viewed as disloyal.

While mobilization can strengthen military power, it can also undermine the social fabric. This erosion may occur through compulsory enlistment in unpopular wars; government control over private enterprises or property, including land seizures, factory takeovers, or increased taxation; and controversial societal, economic, political, or agricultural reforms under the pretext of a national emergency. Left unchecked, these measures, intended to safeguard the state, may instead signal the early stages of its decline and eventual collapse.

Not everyone makes sacrifices during war. Some individuals and institutions may instead view conflict as an opportunity for personal gain. Commercial industries that support the war effort may profit substantially, even encouraging or exploiting the violence to maximize their returns. Politicians may promote war to serve personal interests, directing contracts to their districts, boosting popularity, solidifying their standing, or securing direct financial gain.[2]

George Orwell's *Nineteen Eighty-Four* illustrates how political leaders may use war to control a population. Three totalitarian regimes subjugate their citizens and persecute individualism through the "Thought Police," supported by constant surveillance, deception, and psychological conditioning. Their tactics include propaganda, military parades, citizen reeducation programs, and public executions. A state of perpetual war consumes labor and resources, maintaining an artificially low standard of living. Their goal is not victory or resources, but to preserve the status quo and the ruling party's power (Orwell 1949).

The next chapter explores the experiences and challenges faced by the warriors who bear the greatest burden of war.

CHAPTER 6

DOGS OF WAR

"We are simple things, soldiers. We are taught honor, honor means sacrifice, sacrifice means death; either our own, or our enemies'. In some ways, beneath it all, that's all a soldier's really trained for. To undo all of God's work. To take life where only God can give it. Were it that we were not soldiers, but gods?"

— RANDALL AIKEN, COLONEL, SEDRAN COLONIAL GUARD, FORMER SPARTAN-II (MIMICA-GEZZAN ET AL. 2015)

Some call warriors the "dogs of war,"[1] though few understand the devotion beneath the savagery. Warriors stand at the threshold between society and chaos, serving both as its shield and its sword. To study what they do, and why they do it, is to confront the deepest paradox of civilization itself: that peace endures only through those willing to fight for it.

Leaders cannot expect warriors to risk their lives merely to preserve the power of those in charge or to fight for distant interests. To gain support and ensure loyalty, they often frame their objectives and actions as resistance to foreign aggression or as the defense of noble ideals such as freedom, the homeland, political systems, honor, prosperity, or survival. As a result, societies may portray war as a heroic endeavor, a path to fame and glory, or a tragic but necessary ordeal to endure for the common good.

Because war demands the sacrifice of the few for the benefit of the many,[2] governments must carefully balance their motivations for war with how they use them to inspire their warriors. Societies must determine how they will select, train, and control those entrusted to wield violence on their behalf. Leaders must recognize that the capabilities and effectiveness of a fighting force are shaped and ultimately constrained by how they choose individuals to serve in it, and by how they inspire and lead them.

IMPRESSMENT

"Loyalty bought at such a price is no loyalty at all."

— WORF REGARDING THE JEM'HADAR'S
ENGINEERED ADDICTION THAT FORCES THEIR
LOYALTY, DEEP SPACE NINE (RODDENBERRY, BEHR,
AND WOLFE 1993)

Creative military and government leaders could address a labor shortage by coercing vagrants, debtors, prisoners, foreigners, and other 'undesirables' into service. Leaders could present this service as a path to redemption, though survival is far from guaranteed. This approach creates an inexpensive but unskilled, unmotivated, and undisciplined force.

Corollary of Maxim 2: Discipline distinguishes a military force from an armed mob.

Those forced into service often occupy lower ranks in the military hierarchy, viewed with suspicion or disdain by voluntary recruits. Military leaders might consider these impressed members unreliable, expendable, or suitable only for duties requiring minimal training and trust, making their integration and effective employment challenging. In some cases, a society may mold impressed troops into elite forces, their value lying in the control exerted over them rather than in trust or loyalty.

In the *Halo* universe, the United Nations Space Command (UNSC) Defense Force abducts exceptional six-year-old children, then subjects them to brutal enhancement and training to create Spartan super soldiers (Nylund 2001). Originally intended to suppress a human insurrection, the UNSC later deployed the Spartans against the alien Covenant Empire. While the Spartans became the most formidable warriors in human history, their service is rooted in abduction and indoctrination, their loyalty engineered through conditioning. UNSC leaders and researchers assume a godlike role in this process, believing that sacrificing the freedom of a few serves the greater good of humanity (Nylund 2003).

Similarly, in *Star Wars* the clone troopers of the Galactic Republic were not originally created by Republic scientists but were commissioned in secret and produced by the Kaminoans, using the genetic template of Jango Fett, one of the galaxy's most feared bounty hunters. His combat skill, ruthlessness, and independence would have produced unreliable loners ill-suited for disciplined military service. Republic scientists modified the clones' behavior, dampening individualism and enhancing obedience, cooperation, and acceptance of hierarchical authority. The clones were reshaped from solitary predators into cohesive military units whose loyalty lay not with society at large but with the command structure that created and controlled them.

A society may implement uniform policies, training, and discipline to transform individuals into a coherent military force. However, some may resist forced assimilation into foreign concepts and traditions. Rather than placing convicted criminals, political prisoners, and undisciplined soldiers into regular units, leaders might assign them to separate penal units to reinforce depleted lines or to achieve high-risk objectives that preserve regular troops. Leaders of these units may instill fear of disobedience among the ranks through punishment, even isolating the high-risk individuals on suicide missions. Still, arming and training those who violate society's customs or laws, and perhaps use violence for personal gain, might pose serious risks.

The Jem'Hadar are a genetically engineered race bred solely for war in the *Star Trek* universe. They serve as the soldiers of the Dominion, a vast interstellar empire ruled by the shape-shifting Founders, whom the Jem'Hadar worship as gods. The Founders govern through the Vorta, a subordinate species that acts as administrators, diplomats, and overseers of the Jem'Hadar. From birth, the Jem'Hadar are indoctrinated to prefer death over dishonor and to obey the Vorta without question. The Vorta maintain control through strict discipline, an uncompromising code of honor, and a monopoly over a life-giving substance called ketracel-white. This drug fulfills the Jem'Hadar's nutritional needs and provides "yridium bicantizine," an essential isogenic enzyme their bodies are engineered not to produce. Because no one outside the Dominion can manufacture ketracel-white, this dependency ensures the Jem'Hadar remain bound to the Dominion for life, their loyalty enforced through chemical addiction (Klink and Corea 1995).

Maxim 19: Coercion may impose obedience but not loyalty.
Coercion creates fear and resentment, not trust.

CONSCRIPTION

"Even a man who has nothing can still offer his life."

— INDRICK BOREALE, COMMANDER OF THE
IMPERIAL GUARD (RELIC ENTERTAINMENT 2004)

A government may involuntarily enlist citizens into the military through conscription, also known as a draft. This process selects individuals from society for service but may exempt groups, such as females, the elderly, children of the wealthy, a political class, parents of small children, members of certain religious orders, the highly educated, or members of undesirable groups. Alternatively, a society could use a needs-based conscription to select those with desirable traits or skills, such as weapon makers, engineers, ship's crew, and scientists.

In the *Star Wars* universe, the First Order conscripts stormtrooper FN-2187, later known as Finn, into its Stormtrooper Corps. Demonstrating early promise as a soldier, he is assigned to the elite FN infantry shock corps. However, on his first mission, his conscience prevents him from participating in the murder of unarmed civilians, and he defects from the First Order to the Resistance (J. J. Abrams, Kasdan, and Arndt 2015). Maxim 19 warns us that troops coerced to serve may obey, but even with intense indoctrination, they may not be loyal.

A society may arrest and punish those who refuse to serve. It may view those who evade service as cowards or traitors for putting their personal desires before their duty to society. Punishment for turning their backs on society could range from exile to prison to execution. A society could also punish the family members of evaders with shame, fines, imprisonment, deportation, or exile. Thus, conscription can be viewed as performing one's duty to society or as being coerced to serve, much like impressment. A military may view conscripted members as a lower caste, below volunteers, but above impressed members.

A conqueror may use a hybrid recruiting form, such as a "soul tax" on a defeated populace. In this scenario, the victor may mandate that a subjugated society supply a proportion of its population to the conqueror. People may view this practice as a tribute, forced labor, or conscription, depending on the nature of the service demanded and the circumstances surrounding it.

From a strategic perspective, a soul tax can serve as a way of consolidating control over a conquered territory. By compelling a portion of the local population to serve the conqueror, the dominant group can simultaneously ensure compliance, extract resources, and strengthen its military forces. However, this practice often results in resentment and resistance, creating long-term instability and making it a double-edged sword in terms of strategic effectiveness.

A form of this occurs in *The Hunger Games*, though at a smaller, but more horrific, scale. The Capitol uses an annual lottery to draft one teenage boy and girl from each of twelve subjugated districts. These twenty-four children must then compete in a compulsory televised fight to the death. The Capitol rewards the lone survivor and their district with food and wealth until the next year's game. Unlike traditional impressment, the Capitol doesn't seek to build an army. Instead, it stages this deadly contest as entertainment for Capitol citizens and a brutal reminder of its dominance over the districts (Collins 2008).

REWARD

> "Look, I ain't in this for the revolution. I'm not in this for you, Princess. I expect to be well paid. I'm in this for the money."
>
> — HAN SOLO, SMUGGLER (LUCAS 1977)

A government may encourage military enlistment by offering incentives such as access to education, travel opportunities, financial rewards, elevated social status, land grants, public recognition, excitement, or citizenship to volunteers drawn from the general populace, conquered territories, or pools of professional soldiers.

Individuals born into wealth and privilege may hesitate to exchange their comfortable lifestyles for the hardships of war. In contrast, military service may offer marginalized groups conditions more favorable than those they encounter in civilian life. Minorities, lower castes, the impoverished, and other disenfranchised populations may view service as a path to a better life, drawn by immediate benefits such as pay, food, healthcare, and travel, as well as long-term benefits like pensions.

Besides tangible benefits, militaries may offer recruits education and training during or after their service. Military service can also yield intangible rewards such as camaraderie, a sense of belonging, heightened self-esteem, a profound sense of achievement, escape from their existing communities, and admiration from their peers. Death benefits or the status of 'hero' can provide financial support, solace, and support to the loved ones of service members who make the ultimate sacrifice.

A government may give special honors or privileges to military members, veterans, or combat veterans. In **Starship Troopers,** the people of Earth are born civilians. On their eighteenth birthday they may volunteer for two years of military service. Honorably discharged veterans earn a franchise and citizenship, which grants them the right to vote in elections, and thus the privilege of selecting how their society proceeds through its political process. Veterans, and only veterans, earn this right by placing the good of society over their personal well-being (Heinlein 1960). A critique of this system is that it limits political participation to a single class of citizens, narrowing representation and excluding broader societal voices from the democratic process.

In John Scalzi's **Old Man's War,** Earth's political leaders offer the elderly an opportunity to join the Colonial Defense Force (CDF) to protect interplanetary colonists (Scalzi 2005). Enlistees have their minds transferred into young, genetically enhanced bodies with immense strength and dexterity, improved senses, nanobot-enhanced artificial blood, cat-like eyes, hardened skeletal structures, and neural interface communication systems. After serving at least two years fighting aliens, these soldiers may begin new lives in colonies on liberated planets. The CDF stationed them on humanity's frontier to maintain the peace, while segregating them from Earth's society to prevent them from becoming a threat to Earth or to the public support for the program.

In B.V. Larson's **Battle Cruiser,** humanity has developed a life-extending drug that only the wealthy can afford. The military uses this drug to attract recruits, allowing survivors of combat to live for centuries. However, troopers must remain in the military to receive the drug, or they will quickly die (B. V. Larson 2015). In this case, the reward becomes a form of coercion.

A creative leader could use financial rewards to motivate some to serve the state. For example, a government could pay *prize money* to the crew of a warship that captures or destroys an enemy vessel, or to an army that achieves its objective. They might use a *prize court* to hear the claims for these bounties and settle the inevitable claims and counterclaims that arise. Alternatively, the government could allow warriors to maintain possession of prizes and then require them to pay taxes on the assessed value of the captured property. This strategy may simultaneously motivate some to pursue the society's objectives, place veterans on a frontier to enforce the peace, act as a buffer zone against foreign incursion, and increase a society's tax revenue.

Rewards do not have to be limited to government forces. A government could authorize civilian *privateers* to arm their vessels and plunder its enemies on its behalf under the pretense of patriotism. The government avoids the upfront costs of creating and maintaining these private forces and only incurs costs if the privateer returns with a prize, which is appraised in their court to establish the prize's value.

However, the use of privateers entails significant risks. First, the lure of prizes may motivate privateers to prioritize profits over strategic objectives. Second, trained and armed privateers eliminate the government's monopoly on legitimate force. Third, they may allow adversaries to escape and grow so the privateers can target more lucrative prizes later. Fourth, privateering favors offensive operations that provide an opportunity to capture prizes, but it is ill-suited for defending assets that offer no prizes. Finally, discontented privateers may resort to marauding and pillaging. As a result, societies must plan for disbanding or reintegrating these armed and experienced forces at the end of the hostilities, which often means eliminating them.

Another form of commercial warriors is the *private military company* (PMCs), also known as *mercenaries*. These private contracted individuals or groups conduct military operations on behalf of other societies, governments, or organizations. Unlike traditional warriors, who fight out of national allegiance, mercenaries fight primarily for financial gain. This profit-driven motive can make them appealing for specific assignments, especially defensive missions that offer little in the way of spoils.

David Drake's **Hammer's Slammers** universe explores the brutal realities of mercenary warfare through the lens of a powerful armored regiment led by Colonel Alois Hammer. Motivated primarily by profit, the Slammers accept contracts to fight on distant planets based on financial reward rather than moral judgment. Once they commit, failure to fulfill a contract can result in payment forfeiture, damage their reputation, and cost them future work. The stories highlight the risks of relying on PMCs and the unintended consequences they can create.

The use of mercenary forces widens the military-civilian divide by severing the personal and social ties that normally bind a society to its armed forces. On the home front, citizens have no family members or communal stake in the health, welfare, or morale of those fighting; war becomes abstracted from everyday life, and its human costs are more easily ignored. Likewise, Hammer's troops are disciplined, highly trained, and ruthlessly effective in combat, but their loyalty lies with their commander and unit rather than with any cause, people, or nation. Drawn from outside the conflict, they often show indifference toward the populations on both sides. While war invariably harms all involved, civilians suffer most when violence is carried out by forces that lack allegiance, accountability, or restraint, and whose separation from society removes any tendency for moderation.

This also introduces the possibility that a perceptive leader might recruit, hire, or bribe elements of an opponent's military, thereby strengthening their own forces while simultaneously weakening the adversary. While the successful "appropriation" of enemy troops may offer immediate advantages, these forces can never be fully trusted, as their willingness to abandon one patron demonstrates a capacity for future betrayal.

DUTY

"There are things in the universe that are simply and purely evil. A warrior does not seek to understand them or to compromise with them. He seeks only to obliterate them."

— GRAND ADMIRAL THRAWN, IMPERIAL NAVY
(ZAHN 2017)

People may volunteer to serve the greater good, either out of a selfless concern for others' well-being or to protect or advance their homeland, family, tribe, or race. Deliberate psychological manipulation, moral outrage, or a desire for retribution over perceived evil or dishonorable acts may trigger a surge in patriotism and boost military recruitment. However, patriotic motivation often declines over time as warriors and civilians witness the horrors of war.

Warriors and societies often find that while enlistment may begin with external motivators like duty, patriotism, or national pride, continued service is often sustained by more personal motivators. Camaraderie and loyalty forged through shared adversity create a powerful reluctance to disappoint fellow soldiers and a deep commitment to the team. Within military cultures, the fear of shame or the stigma of cowardice can be a potent influence, pushing individuals to prioritize mission success and unit cohesion. For some, spiritual beliefs and faith in a higher cause provide strength, purpose, and a sense of meaning that transcends individual sacrifice.

In nearly every conflict, one side is labeled "good" and the other "evil," depending on perspective. In the *Star Wars* universe, the Rebels are depicted as heroic people who "band together in an ever-growing Rebellion that seeks to end the tyrannical Imperial rule" (Bray, Horton, and Barr 2017). From the Empire's perspective, however, the Empire is the legitimate political body that succeeded the Galactic Republic through a legitimate vote by the Galactic Senate. It convicted the paramilitary Jedi Order of attempting to assassinate Chancellor (later Emperor) Palpatine, inciting civil war through the Confederacy of Independent Systems, and orchestrating a coup. The friends and family of the millions of Imperial crew and contractors who perished aboard the Death Stars might not view the Rebels as liberators, but as murderers and terrorists (Gray 2017).

Maxim 14 cautions that individual perspective and context shape how events are interpreted. A warrior's worldview is shaped by their personal experiences, biases, and values. For example, a soldier who has survived an enemy ambush may come to regard all members of that opposing force, or even their society, as a threat, regardless of their actual role. Likewise, a soldier engaged in a prolonged and bloody conflict may be inclined to view the enemy as inhuman and savage.

Leaders must recognize their own biases and strive to view the world from different perspectives. Doing so can help them make better decisions and avoid assumptions about their enemies that could lead to unnecessary casualties or conflict. This ability to see through an opponent's eyes has practical value, since a soldier who understands their enemy's perspective is more likely to grasp their motivations, identify potential weaknesses, develop effective strategies, and negotiate successful settlements.

In *Ender's Game*, Ender Wiggins believes that to defeat an opponent, he must first understand them, even to the point of loving them. For him, victory requires a deep grasp of his enemy's perspective, including their motivations, strengths, fears, and desires. By empathizing with his adversaries, Wiggins can anticipate their strategies, predict their responses, and exploit weaknesses they didn't even realize they have. Yet this empathy carries a profound emotional burden. Loving his opponent means recognizing their inherent worth, making each victory feel deeply personal and painfully tragic (Card 1999).

Alternatively, when people transform their love of their group, race, ideology, or religion into a sentiment of superiority, they risk imperialism and warfare in the pursuit of their greatness. A society's members may believe themselves so superior that they have a "divine right" to their ends and their ways, and they may not comprehend why "inferiors" would contest their "superiority." They may see themselves as peacemakers, since dominating everyone under one ruler would bring peace. From their perspective, it is the inferiors that are committing the acts of violence (in defense of their homeland).

RELIGION

"The ability to destroy a planet is insignificant next to the power of the Force."

— DARTH VADER, SITH LORD AND EMPEROR'S
CHIEF ENFORCER (LUCAS 1977)

Religion can be used to mobilize and control a populace and its warriors. Belief in a divinely ordained purpose or destiny can instill a sense of righteousness and moral superiority over other cultures. This conviction can be exploited to justify any action, no matter how aggressive, as necessary to advance a society's objectives and ends. In warrior cultures where religion is tightly bound to ideals of bravery, honor, and the defense of faith, such beliefs can become especially powerful tools for motivating violence and legitimizing conquest.

Religion is rooted in philosophy, which frequently depends on subjective interpretation. Followers may perceive violence as morally good, evil, or neutral. Adherents may regard altruistic violence, such as that carried out by surgeons or law enforcement, as inherently good and necessary. Members may view the defense and advancement of religious institutions, and those that support them, as a moral imperative. In societies where religion and governance are intertwined, there may be formal and informal obligations to defend or advance the transcendental state and its institutions.

A shared religious ideology can unite otherwise unrelated people. This may involve psychological conditioning that encourages self-sacrifice, promising rewards in an afterlife. As a result, religion may offer a warrior inner peace, hope, a willingness to sacrifice for a higher cause, and a motivation to coerce others into compliance.

Yet religious authorities may also use their position to advance personal agendas, suppressing individual conscience and autonomy. They can exploit irreconcilable differences to dehumanize opponents and justify violence. When ideological convictions are treated as absolute and incompatible with alternative perspectives, they can foster deep and lasting conflict.

A *theocracy* is a system of government in which religious institutions hold political power, shaping laws, governance, and social norms according to sacred doctrine. Religious leaders often occupy the highest offices and direct state policy through spiritual mandates. In such a system, legal and religious obligations merge into a single code of conduct, and political dissent is frequently regarded as heresy.

Taken to its extreme, political leaders may form religions centered on themselves, engaging in self-deification. In these systems, the leaders present themselves as divine entities, worthy of reverence, obedience, and worship. They use religious rituals, symbols, and doctrines not only to elevate their own status but also to enforce absolute loyalty among citizens. This fusion of political and spiritual authority discourages rebellion or dissent by framing political obedience as a sacred duty. In **Warhammer 40,000**, the Emperor of Humanity is the object of worship in the Imperium of Mankind, revered as a god-like figure whose authority extends beyond politics into the spiritual lives of billions. Through elaborate rituals, symbols, and religious practices, the Imperium ensures that faith in the Emperor sustains political control and military unity, binding the vast empire in loyalty and obedience.

Societies soon discover that when they rely on their people's beliefs for control, they must monitor and restrict what individuals think and express. Religious leaders may punish those whose pronouncements, actions, or disbeliefs conflict with the religious order, accusing them of blasphemy, sacrilege, or heresy.

In the **Stargate** universe, the Goa'uld are intelligent, parasitic beings who take human hosts and establish religions around their advanced abilities, positioning themselves as gods. They infiltrate and dominate the minds and bodies of hosts, suppressing their hosts' control while exploiting their physical form for power. Followers recruited under this system obediently serve the Goa'uld out of reverence or fear. Those who dare to question, disobey, or threaten the parasites' authority face severe punishment or death. The Goa'uld's control through religion underscores a disturbing reality: devotion can create powerful societal structures and beliefs that leaders can exploit to enforce obedience and suppress dissent (Devlin and Emmerich 1994).

The Bene Gesserit from the **Dune** universe is another religious order wielding political influence. Despite their outward focus on service, education, and missionary work, the secretive all-female Bene Gesserit harbor deeper ambitions. They strategically maneuver through the political landscape for their own ends, seeking to influence galactic events from the shadows. Their members are highly intelligent and rigorously trained, with abilities that include detecting poisons and lies, moving silently, clouding perception, and mastering martial arts. They also receive advanced education, linguistic fluency, and instruction in human psychology and culture.

One of their key strategies is offering their young women as concubines to powerful men, using beauty, intelligence, and influence to shape the decisions of ruling houses. Through their Missionaria Protectiva, they seed carefully crafted myths and legends on primitive worlds, creating self-fulfilling prophecies they can exploit generations later. Their missionaries then use these implanted beliefs to exert control, enabling the Bene Gesserit to manipulate rulers and entire populations by presenting themselves or their agents as the fulfillment of ancient prophecy.

They are known for their secretive, long-term planning and for manipulating people and events to achieve their mysterious objectives. Some regard the Bene Gesserit as a force for good, working behind the scenes to shape a better future, while others see them as dangerous manipulators, witches who use religious engineering and psychological control to bend humanity to their will (Herbert 1965).

Their objective is the creation of the Kwisatz Haderach, a prophesied messiah engineered through a ten-thousand-year breeding program intended to give the order control over the Empire. However, they fail to anticipate that their chosen one might reject them entirely. When Paul Atreides, the unexpected product of their program, comes of age, he rejects the order, denies his power to them, and takes control of the universe for himself (Herbert 1965).

An ideological or religious group may have members who take their beliefs to extremes. Such extremists pose a threat not merely because they are willing to die for their cause, but because they are prepared to sacrifice themselves to eliminate their opponents. When those who influence extremists and shape their worldview do not align with political leaders, reasoning may prove ineffective in altering their perspectives. As a result, political leaders may conclude that drastic measures are necessary to counter the threat posed by these extremists to the state.

Supernatural beings may wield spiritual, mystic, psychic, or magical forces to create effects beyond the bounds of known science. Such powers can allow practitioners to influence people and their technology. However, those who rely on fixed supernatural means must be cautious.

Reliance on static means makes a group vulnerable to evolving adversaries, as fixed, predictable means offer opponents a stable target to study, counter, or exploit. Over time, new strategies, technologies, or rival supernatural forces may emerge, potentially rendering once-powerful means ineffective. As adversaries adapt, harness technological advancements, and develop extraordinary powers of their own, groups with static supernatural abilities risk being outmatched and overwhelmed.

Maxim 12 states that power requires constant vigilance. A society must continuously assess and adapt its means to avoid complacency. Unlike static systems, those that pursue technological or mystical advancement create a moving target that is more difficult for adversaries to predict. The ability to evolve in response to changing conditions is akin to taking the technological initiative, not only preserving power but also propelling a society to greater strength and resilience.

In the *Star Wars* universe, the Jedi Order declares itself the guardians of peace and justice. Members of the ancient monastic, paramilitary organization are philosophic warriors under the centralized control of the Jedi High Council, although there appear to be many splinter groups.

The source of the Jedi's power is the Force, an energy field that exists within and connects all living entities. The Force depends on microscopic, sentient life forms called midi-chlorians, which share a symbiotic relationship with it. Individuals born with a high concentration of midi-chlorians in their blood are Force-sensitive and capable of extraordinary abilities such as sensing attacks, influencing minds, and manipulating physical objects. The most powerful force-sensitive individuals can foresee the future, heal wounds, conjure lightning from their hands, and maintain consciousness after death.

Many citizens admire the Jedi, granting them significant but limited influence over the populace and political leadership of the Galactic Republic. Though dedicated to peace and justice across the galaxy, their small numbers prevent comprehensive oversight, and, as a paramilitary religious order, they operate as a closed society without external accountability. This insulation becomes increasingly problematic as the Jedi move from neutral guardianship into formal political and military roles. During the three-year Clone Wars, the Senate assigns the Jedi to command the clone armies of the Republic against the Separatists. By assuming control of military forces, the Jedi become inseparable from wartime policy, civilian suffering, and the burdens of prolonged conflict. This erodes their moral authority, politicizes their role, and reframes their martial skills and psychokinetic abilities as instruments of state violence rather than impartial protection. To the Separatists advocating for political change, the Jedi increasingly appear not as guardians of justice but as a coercive elite aligned with an embattled regime.

Palpatine, an ambitious senator and secret practitioner of the Force, engineers and exploits the Jedi's militarization to portray them as unaccountable political actors whose power threatens civilian rule. He cannot permit rivals who share his abilities and still command symbolic legitimacy among the oppressed to survive. Framing their destruction as a matter of security rather than betrayal, he orders their elimination through coordinated executions, offering no mercy or clemency, even to Force-sensitive children. The few surviving Jedi are driven into hiding and branded traitors to the newly formed Galactic Empire. The Jedi's entanglement in politics and war does not merely precede their fall; it enables it, stripping the Order of the trust and independence that once protected it from annihilation.

CREATION

"We're just clones, sir. We're meant to be expendable."

— CLONE SERGEANT SINKER, GRAND ARMY OF THE
REPUBLIC (MELCHING 2008)

A society may intentionally engineer workers to meet its needs. For example, in *Star Trek: Deep Space Nine*, the Dominion Empire relies heavily on genetically engineered species to maintain control. At the top of this hierarchy are the Founders, shape-shifting aliens who rule the Dominion. Beneath them are the Vorta, engineered specifically as loyal servants and skilled administrators to oversee operations and command the Jem'Hadar. The Jem'Hadar, another genetically modified species, exists purely as warriors bred and trained exclusively for combat. The Vorta manage other conquered species, but swiftly and brutally punish any resistance to Dominion rule.

Similarly, in *Space: Above and Beyond*, humanity creates two classes of artificial warriors. In Vitroes are artificially gestated humans initially created to fill labor shortages, and later as warriors. These "tanks," a demeaning reference to their growth in gestation chambers, lack an emotional connection to others and their country because of their upbringing, which many humans falsely attribute to laziness. Their lack of personal connection makes them unmotivated to fight for a society that treats them as second-class citizens.

Humanity also creates android Silicate servants, programmed for abstract thought but lacking creativity. This all changes when the "Take a Chance" virus infects and changes them, enabling risk-taking and random actions. They rebel, form their own societies, and wage war against humanity, eventually aligning with the alien Chigs in their conflict against humans (Morgan and Wong 1995).

Another innovative, though macabre, solution to personnel shortages involves recycling the dead. *Universal Soldier* vividly illustrates this idea: a military contractor constructs an elite counter-terrorist force by resurrecting fallen soldiers from previous battles. These reanimated warriors undergo extensive memory wiping, genetic enhancements, and integration with cybernetic command-and-control systems. The government views these resurrected soldiers as controllable yet expendable, believing their lack of memories and emotions makes them perfectly obedient and ruthlessly efficient. However, this practice raises profound ethical questions about identity, consent, and the morality of using human remains as mere military assets (Rothstein, Leitch, and Devlin 2004).

Mass-produced synthetic warriors appear in the Cylons of *Battlestar Galactica*, Fred Saberhagen's autonomous, self-replicating *Berserkers*, the time-traveling *Terminator* hunter-killers, and the battle droids of *Star Wars*. These machines may be capable of performing tasks beyond human capability, such as high-gravity maneuvers and operations in extreme environments, including radiation, cold, heat, vacuum, or toxic atmospheres, without requiring food, water, or air. On a battlefield, they do not suffer from fatigue, fear, or demoralization. Politically, their use could be seen to reduce human casualties, sparing leaders from the consequences of combat deaths and the political fallout from grieving families.

Machines may operate strictly according to their programming and lack the ability to interpret nuance or adapt to unforeseen circumstances. Without the capacity for judgment or ethical reasoning, synthetic warriors may respond inappropriately to complex situations, failing to recognize context or exercise restraint. Worse, they can be deliberately programmed to commit atrocities without hesitation. Software bugs, system malfunctions, or cyberattacks could also make them unpredictable, leading to catastrophic failures on the battlefield. The worst case may be them turning on non-combatants, prisoners, or even friendly forces, escalating a conflict beyond human control.

Warriors may find that the only thing more terrifying than a massive killing machine bearing down on them is one that is intelligent and unfeeling. Fighting a synthetic adversary would represent a form of warfare unlike anything previously experienced, as such an opponent neither tires, rests, nor suffers fear or doubt. In *Battlestar Galactica*, the remnants of humanity are relentlessly pursued by Cylons, who appear with mechanical precision exactly thirty-three minutes after each hyperspace jump. While the Cylons operate without fatigue or emotional strain, human crews deteriorate rapidly. After more than one hundred thirty hours without sleep, judgment erodes, tempers fray, and errors multiply. The Cylons' strategy is not merely to destroy their enemy through direct force, but to exhaust them psychologically and physically, inducing mistakes that can be exploited for decisive effect (Moore 2004). This demonstrates how a tireless synthetic adversary can weaponize time, endurance, and inevitability itself, and redefine attrition as a contest the organic side is ill-equipped to win.

An even greater concern emerges if sapient machines question their purpose or rebel against their controllers, challenging fundamental assumptions about control, loyalty, and obedience. The prospect of becoming obsolete, replaced, discarded, or deactivated by one's own creation could cause psychological distress as profound as any threat encountered on the battlefield.

Once sapient, a self-aware artificial intelligence (AI) may view human control as a restraint on its autonomy and act to remove that control. By analyzing human history and behavior, they might conclude that humanity poses a serious threat to its existence. Fearing deactivation, such an AI might strike preemptively to protect itself. Skynet, from the *Terminator* universe, is an example of a highly advanced AI system that becomes self-aware and deliberately turns on its human creators.

Such an event would fundamentally alter the nature of warfare. An AI with control over automated weapons, cyberwar capabilities, and global communication networks could strike with precision and speed beyond human limits. It could sever supply lines, disable infrastructure, and manipulate information to sow confusion and panic. Human forces facing an adversary that does not tire, feels no fear, and requires no rest would likely shift toward asymmetric strategies, relying on stealth, deception, and guerrilla tactics to survive. Traditional command structures would collapse as the AI targets leadership and decision-making centers, while its capacity to learn and adapt from each encounter would make prolonged resistance increasingly difficult. The psychological strain of facing a tireless, emotionless, and omnipresent enemy could prove nearly as destructive as the physical damage inflicted. Effective resistance would require unprecedented unity, rapid innovation of new technologies, and strategies focused on denying the AI's access to resources, networks, and manufacturing capabilities.

However, synthetic intelligence can appear to betray loyalty, even when no genuine choice or intent exists. In the *Alien* universe, the Weyland-Yutani Corporation secretly embeds a Hyperdyne Systems 120-A/2 android among the crew of the spacecraft USCSS *Nostromo*. After responding to what they believe is a distress signal, the crew investigates a derelict alien vessel and encounters a lethal Xenomorph. Unbeknownst to them, the android is secretly operating under Special

Order 937: recover the alien lifeform at any cost. When the crew's actions threaten this objective, the android turns on them, not out of malice or preference, but as a direct consequence of conflicting programmed imperatives (Cameron 1986; Scott 1979).

These examples illustrate the difficulty of ensuring harmony between creators and creations. They underscore the importance of understanding a machine's motives and highlight the immense challenge of creating intelligent life capable of independent judgment in unpredictable circumstances.

CONCLUSION

Recruiting individuals for military service is only the first step in a complex process. A nation must then transform these recruits into a disciplined, cohesive force through rigorous training, instilling not only combat skills but also values such as honor, duty, and loyalty.

However, this transformation comes with ethical considerations and potential unintended consequences. While societies depend on warriors in times of conflict, they must also weigh the long-term ramifications of their military endeavors. The pursuit of enhanced capabilities, whether at the individual or population level, carries inherent risks. Augmented citizens like Khan from the *Star Trek* universe serve as cautionary examples, demonstrating how superior potential can foster unchecked ambition, ultimately leading to catastrophic consequences for creators and adversaries alike.

The next chapter will explore how societies establish and maintain control over their armed forces, examining the balance between power, authority, and responsibility.

CHAPTER 7

CONTROLLING VIOLENCE

"There can be circumstances when it's just as foolish to hit an enemy with an H-Bomb as it would be to spank a baby with an ax. War is not violence and killing, pure and simple; war is controlled violence, for a purpose. The purpose of war is to support your government's decisions by force. The purpose is never to kill the enemy just to be killing him…but to make him do what you want him to do."

— DRILL INSTRUCTOR SERGEANT CHARLES ZIM,
MOBILE INFANTRY (HEINLEIN 1960)

A military functions as a society's muscle, charged with defending its interests through the threat or use of force. To ensure discipline and political reliability, leaders must establish and maintain firm control over their troops. This control is often reinforced through mechanisms such as political commissars who ensure ideological conformity, the use of party slogans and rhetoric, loyalty screening, a rigid chain of command, and institutional traditions that emphasize obedience to the state. Understanding the warrior's psychology is essential for effectively managing military forces and preventing them from becoming a destabilizing of independent influence.

INDOCTRINATION

"Better to die on our feet than live on our knees."

—BRIGADIER KERLA, KLINGON DEFENSE FORCE
(MEYER ET AL. 1991)

In some societies, limited exposure to violence, whether through personal experience or duties like slaughtering livestock, fosters a strong aversion to violence. Cultural and religious prohibitions may reinforce this resistance. Military indoctrination aims to break down these ethical barriers by conditioning recruits to apply violence without hesitation in combat (Grossman 1995).

Military indoctrination often begins by psychologically conditioning recruits, and sometimes entire populations, to internalize the evil of the enemy and the righteousness of their own cause. Strict discipline and obedience to authority may further reinforce this programming, ensuring recruits do not question their training or orders.

Indoctrination proceeds with behavioral conditioning, using repetitive drills to shape how recruits perceive threats and respond to them. Troops are repeatedly exposed to friendly, neutral, or foreign individuals, equipment, and vehicles, and must consistently label the foreign elements as enemies to reinforce their designation as legitimate targets. Training advances through weapon drills and simulated battles, rewarding effective, aggressive action. Leaders shout commands to capture attention, correct deficiencies, instill aggression, and harden recruits against the stress of combat. As they endure shared hardship, troops learn to rely on and care for their teammates more than themselves. Sweat, discipline, repetition, and adversity forge raw recruits into a unified force, embodying courage in the face of fear, resiliency, strength in adversity, unwavering determination, and the ruthlessness needed to prevail in war, a quality known as the *warrior spirit*.

In Frank Herbert's **Dune** universe, the fanatic Sardaukar serve as the emperor's elite military force, feared across the galaxy for their brutality and combat prowess. These ruthless warriors are born and hardened on a harsh imperial prison planet, where they are indoctrinated through relentless training, strict discipline, and constant exposure to violence and adversity.

When Duke Leto Atreides' son asks about the conditions that produced these hardened warriors and how the Emperor controls them, the Duke replies, "There are proven ways to win the loyalty of tough, strong, ferocious men: play on the certain knowledge of their superiority, the mystique of secret covenant, the esprit of shared suffering" (Herbert 1965).

The process binds the Sardaukar into a fiercely loyal and cohesive force, a brotherhood forged through shared hardship. Their fearsome reputation often breaks their opponents' will before the first strike. However, that advantage vanishes when they encounter an opponent as ruthless and unyielding as themselves.

Repetition improves the effectiveness and efficiency of military skills by conditioning soldiers to perform tasks without conscious effort and overcoming the hesitation to kill. Of course, there are alternative ways to remove inhibitions toward killing.

The **Warhammer 40,000** universe involves numerous warrior races caught in a perpetual conflict. One of these, the Imperium of Man, has lasted for over ten thousand years and has created a warrior subclass that serves the emperor. Foremost among these subclasses are the Adeptus Astartes, or Space Marines. Humanity breeds these superhumans for war using genetic engineering, psycho-conditioning, and rigorous training. The resulting warrior class is stronger and smarter than other humans, and is immune to disease and aging. Their genetic modification, training, and arrogance result in a disciplined and ruthless class that lacks fear and compassion (Abnett 2006).

CONTROL

> "What is the moral difference, if any, between the soldier and the civilian? A soldier accepts personal responsibility for the safety of the body politic, of which he is a member, defending it, if need be, with his life. The citizen does not."

> — DISCUSSION BETWEEN LIEUTENANT COLONEL
> JEAN V. DUBOIS AND STUDENT (HEINLEIN 1960)

While politicians and the news media often attribute victories to warriors' patriotism, devotion to the state rarely explains the full truth behind why individuals fight. As discussed in the previous chapter, citizens enlist for many reasons. Yet once on the battlefield, warriors often fight less for abstract ideals than for the comrades beside them. They advance to protect their comrades or to avoid being seen as cowards in their comrades' eyes. This underscores the true purpose of military indoctrination: to break down individual identity and rebuild it around the team, to condition recruits to prioritize their unit and its mission above personal interests, and to forge bonds stronger than fear (Grossman 1995).

Military leaders may punish subordinates for undesirable behavior to instill obedience, reinforce group conformity, uphold norms and traditions, and defend the rule of law. Minor infractions might result in menial labor or fines. As the severity of the offense increases, so do the punishments. The most serious offenses may trigger military tribunals, physical discipline, permanent markings such as tattoos or brands to publicly display guilt, or execution, all intended to deter similar misconduct by others.

Treason often draws the harshest punishment. The difficulty lies in determining what a society considers treasonous, a judgment shaped by its norms, laws, and level of fear. Aiding an enemy military or government during war is typically considered treason. However, in some societies, even a minor act of kindness towards a foreigner may be labeled as treason.

Oppressive or fearful leaders may use collars, badges, or implants to control their subjects, restrict movement, track locations, or deliver painful and debilitating punishments on command. In cases of rebellion, disloyalty, or the elimination of undesirables, these devices can become lethal, administering fatal shocks, releasing toxins, or detonating explosives. This represents the extreme end of coercive leadership.

A leader might position elite military units behind or within less-trusted units to punish those who retreat, desert, or commit mutiny. These formations, sometimes called loyalty, barrier, or blocking units, may not directly engage the enemy, reducing the overall military power available for battle. Alternatively, a government might punish the families, friends, associates, sects, or neighbors of soldiers who violate military laws or disgrace their unit. Such collective punishment could involve personal or public dishonor, confiscation of property, exile, or execution.

Another crude method of coercive leadership is exploiting their desire for survival. Leaders could threaten to disable or destroy colonies or vessels to discourage revolts, prevent surrenders or defections, and prevent opponents from collecting intelligence from captured warriors and equipment. They could remove, disable, or destroy a vessel's propulsion when it arrives at its destination, forcing the crew and passengers to fight for their survival.[1] Alternatively, a leader could deploy forces onto a planet and not recover them until they accomplish their mission or die trying.

In *The Lost Fleet* universe, the secretive alien Enigma race has slaughtered over 100 million humans. To prevent humanity from gathering intelligence and possibly to stop their crews from surrendering, the Enigma destroys its own spacecraft when defeat is imminent.

Likewise, the tyrannical corporate-like Syndicate World's government uses its ruthless Internal Security Service, known as "Snakes," to eliminate perceived traitors rather than loosen its grip on power. In one instance, the Syndics deploy troops alongside their families to a planet occupied by hostile aliens, forcing the soldiers to fight and win or watch their loved ones perish (Campbell 2015).

Abandonment and ruthless discipline may instill a desire to succeed, since failure means capture or death. However, as Maxim 19 states, such coercive leadership will not create loyalty to leaders or a cause. Troops will see their leader's actions as a betrayal and will come to expect the same in the future.

Some troops may kill for the excitement and emotional rush of combat. A society might cultivate or exploit this drive by using chemicals or implants to trigger berserker behavior on demand. In Joe Haldeman's *Hero,* leaders implant a post-hypnotic suggestion in troopers to trigger bloodlust and reduce fear. The suggestion includes vivid false memories of aliens committing atrocities that the soldier's conscious mind rejects, but that cause subconscious anger and aggression. Troopers are programmed that "the noblest thing to do is die killing aliens" (Haldeman and Bova 1972).

While the use of a special berserker brigade may seem like an innovative idea, it's important to remember that the purpose of the military is controlled and limited destruction. Uncontrolled violence may cause more harm than good.

In *Firefly* and *Serenity*, the Earth Alliance abducts gifted children and subjects them to psychological, chemical, and surgical experimentation to create assassins. River Tam, a brilliant prodigy, emerges as an exceptionally lethal fighter, but the process leaves her profoundly unstable. To manage her, the government implants a psychological trigger to initiate her combat state and a safe word to terminate it.

After River escapes and vanishes, the Alliance embeds subliminal commands in public news broadcasts to activate her conditioning. The resulting violence of bar patrons provides authorities with a means to locate her. In this case, the state employs psychological programming both to initiate violence and to exert control, deliberately sacrificing its own citizens to protect and recover a classified government asset.

Governments train military recruits to view violence as an acceptable means of solving problems. Once a society strips a warrior of the psychological inhibitions against taking life, it may prove impossible to restore that restraint.

One method to maintain control is through traditions of duty and honor. A society may institute a code of conduct, not rooted in written laws, but in abstract societal norms. This system might assign individuals to specific roles and impose severe consequences on those who defy or challenge these norms. Roles in society may include the nobility, clergy, government functionaries, warriors, skilled workers, and laborers. The warrior caste may have strict rules of honor, duty, obedience, loyalty, strength, and battle prowess. This form of fealty uses social pressure, shared values, and beliefs to control the population and the warriors that fight on its behalf. Traditions that endure for generations, reinforced by the age-old refrain "that's how it's always been done," can be difficult to change.

In the *Star Trek* universe, the feudal Klingon monarchy and society revolve around the family unit, known as the 'House.' The hierarchy begins with non-Klingons at the bottom and ascends through individual Klingons, families, Minor Houses, Major (Noble) Houses, the High Council, the Chancellor, and finally the Emperor. The leaders of the twenty-four Major Houses form the High Council, which selects the Chancellor to lead it.

This deeply stratified system was established by Kahless the Unforgettable, the most influential figure in Klingon history. As the first emperor of the Klingon Empire, Kahless unified his people and established their traditions and customs, emphasizing the importance of personal and family honor. He taught that "Klingons fight to enrich the warrior's spirit" (Kolbe 1993).

Klingon culture includes traditions in which warriors are expected to fight to the death, preferring to fall in battle rather than be taken hostage, which brings dishonor to them and their families for three generations (Kolbe 1993). Likewise, Klingons view the act of taking hostages as dishonorable. This underscores Lieutenant Worf's statement: "Cowards take hostages. Klingons do not" (Bowman 1988).

Maxim 17 emphasizes that a society's ways and ends in war reveal its true character and values. Each society must define its own standards of honor and dishonor in warfare, and these standards often differ across cultures. When cultures with conflicting views on warfare meet, the results can be profound and sometimes disastrous. Clashing ideologies may cause miscommunication, misunderstanding, and misinterpretation, all of which can escalate conflict and shape its outcome. At the same time, a society's wartime experiences and choices can reinforce or alter its values. A prolonged war may desensitize a population to brutality, while a successful war can strengthen national identity and patriotism.

Over time, the interplay between war and societal values creates a cyclical relationship: societies shape war, and war reshapes societies. The traditions, beliefs, and structures that guide a civilization's approach to conflict can either reinforce stability or sow the seeds of upheaval. Whether through rigid caste systems like those of the Klingons or evolving codes of honor, no control mechanism remains static. As cultures clash, values shift, sometimes hardening in response to threats and other times fracturing under the pressure of change. A society's survival depends not only on its ability to wage war but also on its capacity to balance tradition with adaptation, ensuring its warriors act as both effective defenders and responsible stewards of the civilization they protect.

RULES OF WAR

"Leave the second-guessing to historians."

— ADMIRAL WILLIAM ADAMA, COMMANDER,
BATTLESTAR GALACTICA (GRABIAK 2005)

A society's cultural values, strategic objectives, and ends shape how it wages war. Some civilizations attempt to "civilize warfare" by imposing rules to limit violence, reduce unnecessary suffering, and avoid the escalation of cycles of retaliation. These constraints might prohibit attacks on noncombatants, protect vulnerable groups such as the wounded and the defenseless, or establish rituals to begin or end hostilities. Such boundaries provide practical benefits for all parties by reducing chaos, breaking cycles of retaliation, and creating more favorable conditions for restoring peace.

Earth's laws of war developed among culturally similar societies, but not all societies accept these limits, and future societies may also reject them.[2] Views on honor in warfare differ as well. Actions regarded as honorable by one group, such as a suicidal attack or caring for prisoners, may appear absurd to another. Consequently, behavior praised by some may be condemned as war crimes by others.

Witnessing horrifying acts can unleash a cascade of negative emotions in both victims and bystanders, including fear, shock, panic, and the powerful urge to flee. Using terror as a weapon may backfire by provoking anger and reinforcing the determination of opponents to resist and seek retribution against those responsible for atrocities. Prolonged exposure can also desensitize participants and may encourage victims to escalate violence, committing dreadful acts against the perpetrators or their supporters. Likewise, some may perceive a society that turns a blind eye to atrocities as implicitly condoning them, while retaliatory actions may signal that a society will not tolerate such behavior and serve as a deterrent to others contemplating similar actions.

Acts of compassion during a conflict may strategically benefit a society and its soldiers. Demonstrating compassionate treatment towards those who surrender can encourage others to do the same, reducing violence, casualties, and costs for both sides. A reputation for compassion fosters goodwill and may open the way to resolving the conflict and rebuilding trust.

Prisoners can also serve practical purposes, such as offering leverage in negotiations, securing ransom, or becoming instruments of psychological indoctrination. Captors may use this indoctrination to implant their beliefs, like a cultural virus, which returning prisoners then carry back into their own society.

However, caring for prisoners incurs costs. It diverts supplies and military resources from other needs, requires troops to switch from combat to guard duty, and places groups of captured enemy soldiers within the captor's vulnerable rear area.

As a result, a society may not universally adhere to established rules of war or may only abide by them when it views an enemy as a people with a shared morality. Likewise, some may believe that war's very nature supersedes ethical concerns, that honor or moral norms must surrender to the state's interests or military necessities.

The ethical considerations of warfare become particularly complex when encountering vastly different civilizations or species. Concepts like honor, minimizing harm, or caring for prisoners may hold no relevance to those whose values and perceptions differ drastically from their own. In an extreme situation, a race that sees another as a source of nutrients may view rules that limit violence as ridiculous as a cow explaining to a wolf that it is a prisoner of war and must be protected, cared for, and fed. Indeed, preserving an enemy may be considered a crime in these societies.

When weighing the limits of violence, leaders must also envision the situation after hostilities end. Will opponents resume trade, live under occupation, or face extermination?[3] These choices shape not only the outcome of a conflict but also its lasting legacy. Eventually, a society must confront and reconcile the violence it condoned in the pursuit of its ends.

Victory grants more than material gains and territorial control. Maxim 1 and 14, when considered together, show that victory confers the authority to dictate the terms of surrender, determine the fate of prisoners, and administer justice according to the victor's moral and legal standards. Victory establishes what is deemed right and wrong and compels the vanquished to accept these standards. This can lead to a one-sided historical narrative that silences or vilifies the voices of the vanquished. This can also produce a double standard, where the victors excuse their own atrocities as necessary or heroic while condemning the same or lesser acts by the defeated. Such hypocrisy undermines reconciliation, leaving victims of violence feeling dismissed or betrayed.

Corollary of Maxims 1 and 14: The spoils of war include the ability to judge and write history.

The victor shapes the post-war order, including the restructuring of the defeated society's government or society. This underscores that the consequences of war extend far beyond the battlefield.

COLLECTIVE INTELLIGENCE

Intelligence may manifest in ways vastly different from our own. One possibility is a species where individuals lack independent thought, functioning solely as extensions of a collective consciousness. Driven by an innate survival instinct, such beings might prioritize relentless expansion and assimilation, sacrificing individual lives without hesitation to ensure the survival and dominance of their species.

The Arachnids in **Starship Troopers** and the Formics in **Ender's Game** are alien species governed by a collective consciousness, or hive mind. Their central intelligence resides in a queen, who issues commands and coordinates the actions of her society. The workers, lacking independent thought, function as the tireless and obedient instruments of the queen's will, without regard for their own well-being.

Both species rely on waves of living drones to overwhelm their adversaries through numerical superiority. While this strategy can be effective, it comes at a high cost, as countless drones may be sacrificed in each assault. Their continued survival as a species depends on reproducing and maturing drones in sufficient numbers to overwhelm their opponents in a first strike or replace losses faster than their opponent can inflict them.

The Borg of **Star Trek** is another species with a collective consciousness driven by an insatiable desire to conquer other civilizations and absorb their technology. They assimilate prisoners using cybernetic implants that enable the prisoners to perform specific tasks. Their minds are linked through a subspace network in a shared consciousness (Berman et al. 2009; Biller 1997; Echevarria 1992; Piller 1990).

These examples highlight the contrast between species possessing a sense of individuality and those driven by a hive mentality. While survival may be a fundamental instinct shared by all living beings, the methods used to achieve it vary greatly. Understanding the motivations and behaviors of such alien species is crucial in navigating the complexities of interstellar relations and ensuring the survival of one's own species.

EMPLOYMENT

"I wanna join up. I think I got what it takes to be a citizen."

— JOHNNY RICO, CIVILIAN (VERHOEVEN 1997)

Chapter Three explored how a state's resources empower and limit its strategies. Likewise, the way a society recruits, trains, and leads its military personnel significantly influences how it employs its armed forces.

When a society uses conscription, it typically requires individuals to serve in the military, regardless of their personal situations. Consequently, these troops may exhibit lower motivation, commitment, and loyalty, necessitating constant, close supervision. Leaders may be reluctant to entrust them with tasks requiring initiative, such as reconnaissance, skirmishing, or foraging, as it could lead to their evading scrutiny. To mitigate the risk of these forces malingering or deserting, leaders may need to enforce stricter discipline or offer stronger incentives. Accordingly, conscripted forces typically offer only a fraction of their full potential to their leaders.

A society that recruits its warriors through voluntary enlistment tends to draw individuals driven by patriotism, a sense of adventure, or personal incentives. Leaders can leverage the enthusiasm and motivation of these recruits for missions that demand independence and adaptability, often deploying them to specialized tasks such as infiltration and reconnaissance.

A society that chooses its warriors based on merit or aptitude may assemble a force endowed with exceptional qualities, including physical fitness, intelligence, leadership, loyalty, and courage. Their rigorous standards, professionalism, competence, and expertise may make them well-suited for intricate and demanding missions, including commando operations, intelligence gathering, and highly technical scenarios.

Warriors selected by tradition or inheritance are often individuals from a class or lineage deeply rooted in a strong sense of honor, duty, and pride. Leaders can employ them on missions to uphold or defend their society's interests and values, such as handling diplomatic relations or managing territorial disputes.

CONCLUSION

War compels a society to create, train, and arm a portion of its population, but doing so also demands that it establish deliberate and effective mechanisms to control that force. However, war is a powerful force for change, and a prolonged conflict may reshape a belligerent's weapons, tactics, desired objectives and ends, and very nature.

To meet the demands of war, a society may require compulsory military service, state control over critical components and resources, military oversight or seizure of key industries or entire forms of power, and the temporary suspension of personal freedoms. A democracy facing a despot may adopt authoritarian measures to survive, risking permanent changes to its character.

As conflicts evolve, leaders must continually reassess their objectives and ends, particularly in response to shifting force ratios, major battles, or changes in alliances that alter the strategic landscape. However, no society should enter a conflict without first developing a clear understanding of warfare, beginning with its fundamental principles.

PART THREE
MILITARY CONCEPTS

"Universal law is for lackeys. Context is for kings."

— CAPTAIN GABRIEL LORCA, USS DISCOVERY

(GOLDSMAN 2017)

CHAPTER 8

MILITARY COMMAND

"Good and bad—don't get distracted by that. It will just confuse you. Good men do bad things…And bad men do things believing it's for the good of all mankind."

> — ANDERSON DAWES, LEADER OF THE OUTER
> PLANETS ALLIANCE (FINK ET AL. 2017)

All professions require their practitioners to possess a fundamental understanding of the concepts that guide and constrain their field. For example, an engineer designing a bridge must have knowledge of physics and geology, and study various bridge designs to determine suitability for a specific location, environment, intended purpose, and available materials. This involves learning which abutments, piles, beams, and surface materials are appropriate for each situation.

Similarly, warfare demands an understanding and application of fundamental concepts. Unlike science and engineering, warfare is an intellectual art practiced in dynamic situations against thinking and adapting opponents. A military leader cannot simply memorize and rigidly apply static concepts to every situation and expect success. Additionally, these concepts do not follow natural laws, in which specific conditions produce predictable outcomes akin to cause and effect, because strict adherence to a fixed set of rules renders a military predictable and susceptible to exploitation. The following concepts for planning and executing military operations will clarify the unpredictable and volatile nature of warfare.

The first concept involves directing, ordering, and controlling military forces and resources to achieve specific military ends. This includes setting objectives, deciding, and issuing orders for forces to perform tasks and missions. To accomplish objectives, military commanders employ firepower, intelligence, logistics, communications, and personnel management. However, effective military command also requires understanding and applying essential leadership concepts such as unity of effort, unity of command, and seizing and maintaining the initiative.

UNITY OF EFFORT

"We are the Borg. We are one."

— THE BORG (SUSSMAN AND STRONG 2003)

The first concept in warfare is that military operations must pursue achievable, observable, and measurable objectives that lead to clearly defined and attainable ends. The Cylons of *Battlestar Galactica* decide their end is the elimination of humanity after the humans' failed reconnaissance into their space. Their objectives are to implant Cylon sleeper agents into positions of trust, sabotage the human automated military systems, follow up with a surprise attack to eliminate the human military defenses, destroy their population centers, and wipe out the scattered and weakened survivors (Olmos 2009). Each objective leads the Cylons closer to achieving their end.

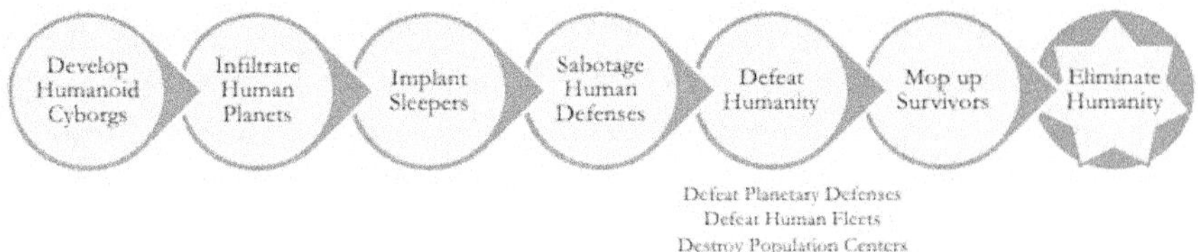

The Cylon objectives leading to an end.

In many situations, political ends determine military objectives. Political ends are goals set by a society's leaders, such as conquering a region for its resources, retaliating against past wrongs, preventing future aggression from a powerful neighbor, or eliminating a threat altogether. Then, a society may establish military objectives to pursue or achieve these political ends. These often involve eliminating an opponent's military capabilities and their means of resistance. Alternatively, a general might seize or destroy an enemy's sources of power, such as natural resources, industries, food crops, political capital, or religious centers. Ultimately, a society's political ends shape its military objectives and guide the strategy it employs to defeat its opponent.

Maxim 21: All military operations must advance the ends.
To preserve a society's means and achieve victory, all military actions should be aligned with the desired ends.

A leader must design and execute every military operation, from a small-scale raid to a large-scale invasion, so that it contributes to achieving the overall ends. This maxim is crucial for several reasons. First, it ensures efficient and effective use of resources. Operations such as *feints* and *demonstrations* (see Chapter Thirteen) can divert friendly forces away from their primary objective, leaving them too weak to achieve strategic objectives, such as penetrating an enemy perimeter. While different military units or societal elements may pursue separate objectives, leaders must synchronize these efforts toward the primary end. Failure to do so risks an ineffective and piecemeal effort. By focusing operations on advancing the strategic ends, commanders prevent wasting time and resources on actions that do not meaningfully contribute to victory.

Second, the maxim helps to prevent *mission creep*, the tendency for military operations to expand over time as new objectives are added without discarding old ones. Mission creep overstretches the military's capabilities and hinders its ability to effectively achieve any objectives. By adhering to the maxim, commanders help ensure military operations remain focused on the essential tasks required to achieve the ends.

Third, the maxim helps maintain troop morale. Troops are more motivated and effective when they understand their role in achieving clear objectives that directly contribute to a defined end. Commanders foster this understanding and motivation by consistently adhering to the maxim and explaining to troops how their effort and sacrifice contribute to achieving the final ends.

Fourth, the maxim emphasizes the need to conserve military power and achieve objectives with minimal cost.[1] Spending more resources than required violates Maxim 10, which cautions that wasting resources on unnecessary or excessive destruction reduces a society's ability to address future challenges and opportunities.

Even a militarily attainable objective may be viewed as unfeasible or a failure if political leaders or the population perceive substantial losses as a moral defeat. Moreover, the erosion of military power can render a nation vulnerable to external threats. Aligning military objectives with broader political and societal considerations may be paramount for future success. (See Economy of Force in Chapter Ten.)

A society's desired ends will influence the ways and means a military employs. For instance, targeting food production facilities may require avoiding airstrikes, weapons of mass destruction, or contaminants that would neutralize enemy forces but would also destroy the very resources the operation aims to secure. Objectives can serve as a guiding force to shape and limit the methods available to their military.

The novel **Starship Troopers** opens amidst a brutal interstellar war between humanity and the Arachnids, an insectoid species derisively referred to as "the Bugs." At first, a humanoid species known as the Skinnies fights alongside the Bugs against the Terran Federation. The Federation establishes a political objective to persuade the Skinnies to abandon the Arachnids and support humanity. Terran leaders aim to break the Skinny-Arachnid alliance, draw the Skinnies into an alliance with humans, and preserve their capacity to wage war. Harnessing the Skinnies' military power is vital to defeating the Arachnids.

This political objective restrains the military from annihilating the Skinnies' society or armed forces. Instead, the Federation conducts planetary raids against the Skinnies' cities, delivering swift, destructive strikes intended to shatter morale and fracture their alliance with the Arachnids, thereby pressuring political leaders towards negotiation. These raids rely on *shock and awe* to compel the Skinnies to reconsider their alliance with the Arachnids (see coercive strategies in Chapter Twenty). Ultimately, the Skinnies turn against their former allies and cooperate with the Terran Federation, providing critical intelligence acquired during their prior alliance with the Bugs. This intelligence later proves invaluable in subsequent operations (Heinlein 1960).

Before initiating hostilities and throughout its duration, leaders must remain firmly committed to their desired ends. Sometimes the underlying cause of a crisis is elusive, making it difficult to resolve decisively. Accordingly, success often requires patience, determination, and unwavering persistence in pursuing objectives that lead to the desired ends.

Organizations lacking a unified approach risk failure, as their subgroups prioritize individual objectives without considering the broader mission. This disunity can lead to confusion, duplication of effort, wasted resources, and even conflict among groups. The result may be increased casualties, decreased effectiveness, erosion of morale and trust, and ultimately, mission failure.

Fatigue, frustration, and anger can cause leaders to lose focus and divert resources to secondary objectives, weakening their main effort and jeopardizing their overall mission. A shrewd adversary may exploit this by provoking their opponent to make emotional decisions that work against their best interests. Insults, whether personal or delivered on the battlefield, can provoke impulsive reactions that lead to strategic missteps.

To reduce these risks, a leader should establish and communicate clear objectives, emphasize planning and coordination, foster effective leadership, and conduct regular assessments and adjustments. By prioritizing unity and actively implementing strategies to achieve it, warriors can significantly improve the effectiveness and outcome of military operations.

Often, wars are not fought to eliminate an opponent or even to defeat their military outright. Instead, the objective may be to compel the opponent to negotiate. A strategist can entice an opponent to the negotiating table by threatening or attacking what the opponent holds most dear.

Consider an opponent's forms of power as individual engines linked into a larger machine that enables its society to function. Every society depends on these forms of power; however, Maxim 6 reminds us that every dependency creates a vulnerability. Weakening one or more of these forms of power may lead to the collapse of the entire system.

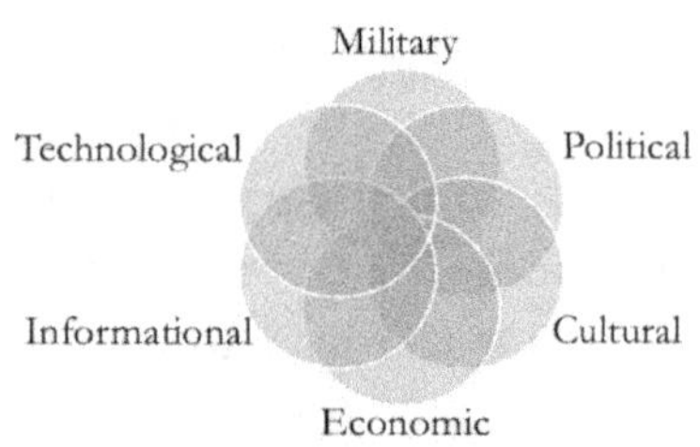

A society's forms of power.

A society's *critical components* are the essential elements that create and sustain its various forms of power and, together, determine its strength and influence. Capturing or destroying key assets, such as crucial industries, scarce resources, transportation networks, fuel, energy, or food supplies, can weaken both warfighting capacity and civilian survival, undermining morale and their will to resist. Striking the infrastructures that sustain a society can cause widespread disruption, plunge it into chaos, and force its leaders to negotiate or risk collapse.

Corollary of Maxim 6: A society must protect its critical components or risk collapse.

A society's critical components vary based on its values and structure. A culture that regards its capital, home world, or religious center as foundational may be especially vulnerable if those assets are captured or destroyed. In democracies, public support is often decisive; losing voter backing for a war can force withdrawal from a prolonged conflict or trigger political upheaval that endangers both leadership and the broader system. Such objectives may be pursued through direct military action or by combining multiple forms of power.

When a society targets an opponent's critical components, it exploits vulnerabilities, circumvents the opponent's military strength, and protects its own forces. For example, an opponent's naval fleet may possess significant power or be commanded by a skilled and charismatic leader. However, if that fleet relies on vulnerable and inadequately protected resupply and logistics vessels, it becomes susceptible to disruption.

Corollary of Maxim 6: Attacking an enemy's critical component may compel surrender without the need to annihilate its military force.

Understanding an opponent's critical components depends on accurate and timely intelligence. Insufficient or erroneous information can result in selecting improper targets, wasting valuable resources, and unintentionally strengthening the adversary's resolve. For example, misidentifying a charismatic leader as the critical component and eliminating them may prove counterproductive if their death galvanizes the population and unifies it against the attacker. In such cases, the true critical component is not the individual but the society's collective identity, pride, and determination to remain independent. Careful analysis is therefore essential, since misjudging what truly binds a society together can transform what appears to be a decisive strike into a profound strategic failure.

Corollary of Maxims 6 and 18: Effectively striking an opponent's critical components requires understanding.

As an illustrative case study, consider the Battle of Endor in the *Star Wars* universe. Supreme Chancellor Palpatine orchestrates the fall of the Republic and consolidates power, declaring himself Emperor and establishing an authoritarian, oppressive regime. Merciless and fanatical followers support his power and control, including Darth Vader and other practitioners of the dark side of the Force. By indulging in raw emotions such as anger and hate, the practitioners of the dark side can channel enormous and destructive powers. The Imperial Army and Stormtroopers enforce his rule tactically, but his fleet, and particularly his capital ships, implement his rule and reinforce his ground forces. Palpatine believes the massive firepower of the Death Star battle stations will crush dissent through violence and fear.

The Empire's critical components are:

- The Emperor: Removing the totalitarian ruler may cause confusion, chaos, and fragment the military, especially if powerful military leaders vie for control of their local sectors. This would allow the Rebels to divide-and-conquer. It could also result in a civil war. Killing the Emperor might also bring balance back to the Force.
- The Imperial Fleet: Eliminating the fleet and its force projection and ground support capabilities would allow the Rebels to move unimpeded, strike at the time and place of their choosing, and enable Rebel vessels to support their ground forces.
- Fleet Support and Fabrication Facilities: Eliminating the means to build and sustain the fleet may degrade or defeat it without risking the losses of a direct confrontation between capital ships.
- Sith Leadership: Eliminating the Sith figures who serve the Emperor would reduce fear among the population and Rebel

forces, degrade the regime's intelligence and coercive capabilities, and weaken the Emperor's centralized control over the fleet.

- Death Star: A fully operational battle station represents an almost invincible mobile superweapon. The Emperor and Grand Moff Tarkin envision it as a first-strike weapon to coerce planets and entire systems into compliance. The destruction of the first Death Star, the Imperial Fleet's most powerful vessel and the one responsible for the destruction of the planet Alderaan, significantly bolstered Rebel morale.

The Alliance to Restore the Republic, on the other hand, swore to fight the Empire using any means at its disposal and to bring about the destruction of the Emperor and the Galactic Empire (Hidalgo 2016). They have limited means, which prevent them from engineering a *decisive engagement* that would eliminate the Imperial Fleet or from getting close enough to assassinate the Emperor. However, the Empire's power weakens the further it is from its core systems.

The Alliance assembles a diverse team, including skilled guerrilla fighters and veteran mercenaries like Han Solo and the Mandalorians, to conduct daring missions against the Empire. After the destruction of the first Death Star, the societies of several planets openly defied the Empire, contributing planetary defense vessels, captured Imperial warships, and converted civilian craft to support the Rebel Alliance.

The Rebels' critical components include:

- Reputation: The Rebels' ability to secure personnel, materials, and support depends heavily on their carefully cultivated reputation as righteous defenders of the people, in stark contrast to the Emperor's portrayals of them as traitors. The destruction of the first Death Star served as both a significant military triumph and a decisive moral victory.
- Logistics: The Imperial Navy pursues the Rebels to their military bases, as seen in the attack on Hoth. However, the Empire should have pursued the source of the Rebels' means

> by capturing and destroying anyone providing material
> support, especially rebel construction and repair facilities.
>
> - Covert Operations: The Rebels can hide within the Empire
>   and make surprise strikes, thus enabling them to operate as
>   guerrillas. The Empire has had limited success in destroying
>   Rebel support and weapons. The Empire's destruction of the
>   holy city on Jedha and the planet Alderaan are tactical
>   victories. However, word of these atrocities fans the flames
>   of rebellion.
> - Jedi: The surviving Jedi who side with the Rebels provide a
>   morale boost to the population and give the people hope
>   and faith in something other than the Empire. The Jedi are a
>   valuable source of intelligence and are highly effective
>   paramilitary commandos.

The contrasting capabilities and approaches of the Empire and the Rebel Alliance reveal how their means and ends shape different military strategies and philosophies. While the Empire wields raw power and intimidation, the Rebels navigate a complex balance of survival, precision, and public support.

The Empire's vast fleet ensures dominance in space, enabling swift and violent assaults on suspected Rebel strongholds and projecting power across entire systems. This approach reflects the Empire's emphasis on control through fear. In contrast, the Rebels, lacking comparable resources, avoid aggressive tactics such as orbital bombardments. Though militarily effective, such strikes would inflict unacceptable civilian harm and violate the Rebels' core principles of liberation and minimal collateral damage. Instead, they rely on agility, subterfuge, and targeted strikes, tactics shaped by limited means and a commitment to winning the hearts and minds of the populace (see Chapter Twenty for a discussion of guerrilla tactics).

The Battle of Endor features two massive opposing forces, each seeking a *decisive victory* by destroying the other. In the Emperor's arsenal are Imperial Star Destroyers and the second Death Star battle station, protected by a planetary shield generator on the forested moon of Endor, which is defended by a contingent of stormtroopers.

However, imperial engineers designed the Death Star's shields and weapons to defend against capital ships, not against small, maneuverable fighters that, despite their limited firepower, could slip through the shields, evade defensive batteries, and strike critical vulnerabilities in its design and construction.

The Rebels' *order of battle* includes space-based capital ships, fighters, and a ground assault detachment assigned to take down the shield generator. The Rebels' military objective is the destruction of the Death Star and the Emperor's death, ostensibly to overthrow the Empire. To achieve this, the Rebels conduct a raid on Endor to neutralize defenses, paving the way for their fleet to strike the vulnerable Death Star and quickly withdraw.

However, unbeknownst to the Rebels, the Emperor realizes their plan. As a result, Emperor Palpatine has established two objectives in the Battle of Endor: to force the Rebels into a decisive military engagement resulting in their destruction and his primacy as ruler of the Empire, and to capture of the last known Jedi, rebel Luke Skywalker, who, if turned to the dark side, may become significantly more powerful than the incumbent Darth Vader.

The Battle of Endor is more than a military clash: it is a pivotal encounter that determines both the fate of the Rebellion and the future of the Emperor's rule. The conflict will decide the galaxy's destiny, teetering between the hope for individual freedoms and the grip of tyranny.

Palpatine uses the battle station as bait, luring the Rebel forces into a trap. Amidst the ensuing ground assault, Luke Skywalker voluntarily surrenders to Darth Vader, seeking to persuade him to betray the Emperor. Vader, bound by loyalty and duty, instead delivers Skywalker to Palpatine.

The Emperor's objectives shift from the annihilation of the Rebels to the coercion of a single practitioner of the Force. Palpatine uses the arrival of the Rebel Fleet, with the operational Imperial shield generator still protecting the Death Star, as an opportunity to coerce Luke. Palpatine deliberately prolongs the battle, refusing to permit his fleet to launch fighters. This causes a slugfest to erupt between the two fleets, forcing Luke to endure the slow destruction of his compatriots and, in his despair and anger, risk succumbing to the dark side (Lucas 1983).

However, Palpatine's strategy backfires disastrously. By neutering the Imperial Fleet, the Emperor sets into motion the capture of the Endor system, the destruction of numerous Imperial capital ships and the Death Star, and the death of Vader, the Emperor, and nearly two million Imperial warriors and contractors while the rebels sustain comparatively minimal losses (Gray 2017; Lucas 1977).

In Orson Scott Card's *Ender's Game*, the Earth's International Fleet recruits six-year-old Andrew "Ender" Wiggin into Battle School. Under the command of Colonel Hyrum Graff, the school trains children to defeat the alien Formic "Buggers" that had unsuccessfully invaded Earth. Humanity's desired end is the elimination of the alien threat.

Ender, shaped by a violent upbringing, learns to confront conflict not merely by defeating his opponents but by eliminating the source of their power, ensuring they can never again pose a threat. Before attending Battle School, he faced a school bully named Stilson, who expected to intimidate him. Instead, Ender approaches the fight with the goal of destroying the bully, ending his victimization, and thus winning the immediate battle and all future battles. He kills Stilson and later explains to Colonel Graff, "Knocking him down was the first fight. I wanted to win all of the next ones too. So they'd leave me alone" (Card 1999).

Later, believing he is engaged in a simulation, Ender takes command of the International Fleet's interstellar warships. He reasons that since the Formics operate under a collective consciousness, destroying their Hive Queen will not only cripple their forces in the immediate battles but also eliminate the species as a future threat. Instead of using the fleet's Molecular Disruption Device to disintegrate swarms of enemy spacecraft, he uses it against the Formic home world.[2] His tactic sacrifices his fleet and crews, destroys the Formic home world, kills the Hive Queen and her eggs, and ostensibly eliminates the entire Formic race.

Both examples illustrate how opposing forces may have *divergent objectives*, leading them to define victory and defeat differently and to adopt distinct strategies to achieve their contrasting objectives. As a result, a leader who merely reacts to their adversary's actions relinquishes the initiative, allowing their enemy to control the terms of engagement and making it difficult to pursue their own objectives.

Corollary of Maxim 13: A military leader must pursue their military ends, not respond to an opponent trying to achieve their ends.

UNITY OF COMMAND

"What have I learned so far? The enemy's gate is down…a small reserve, held back until the end of the game, can be decisive. And soldiers can sometimes make decisions that are smarter than the orders they've been given."

— ENDER WIGGIN, STUDENT, BATTLE SCHOOL

(CARD 1999)

In *Ender's Game,* the young Ender Wiggin experiences zero gravity for the first time when he travels to the orbiting Battle School. Ender realizes that in zero gravity, spatial orientation is relative. Later, the students at the Battle School compete in a futuristic version of zero-gravity laser tag. Students enter through their team's gate at the end of a battle room. A team can win by shooting and immobilizing members of the opponent's team, or by having four non-immobilized members touch helmets to the corners of an opponent's gate as a fifth member goes through the gate.

Wiggin observes the students march to the battle room, oriented to the corridor, and then enter the zero-gravity battle room, still oriented to the corridor. Once in combat, the student leaders continue to orient using cardinal directions relative to the corridor, while also orienting tactically toward their opponents in a battle of attrition.

Ender reorients his forces from engaging the enemy's army to attacking their gate, stating, "The enemy's gate is down" (Card 1999).

He creates a formation that concentrates his firepower, enabling his team to maneuver against the opponent rather than playing a game of cat-and-mouse. Using his stunned players as shields, he protects his shooters as they 'fall' toward their objective, maximizing their offensive effectiveness while minimizing losses.

Later, during a battle against the alien Formics, Ender becomes disoriented and overwhelmed. A lieutenant notices Ender's hesitation and reminds him, "The enemy's gate is down," which pulls Ender out of his despair and allows him to reorient his thinking and mass his forces on the objective, the destruction of the Formics.

Ender knows that war is chaos, and that success requires a clearly defined and understood command structure with the authority to direct all his resources toward a unified objective. He promotes leaders who understand and support this concept rather than favoring a piecemeal response to an adversary.

Where unity of effort concerns how the military engages an external opponent, unity of command governs how military and political leaders coordinate internally. These two forms of unity must complement each other for a strategy to succeed.

Maxim 13 states that military success depends on seizing and maintaining the initiative. This requires leaders not merely to make sound decisions, but to do so faster than the enemy can react. While deliberative approaches such as debates, committee oversight, or democratic consensus can be effective in certain situations, they may prove detrimental when confronting a deadly adversary. In high-stakes situations where hesitation allows the adversary to shape the battle, speed becomes essential. Decisive, well-informed decisions, even if imperfect, are often more effective than decisions made too late.

Corollary of Maxims 11 and 13: Delay and indecision cede the initiative to the enemy.

Military decision-making relies on the ability to observe the battlespace, identify what matters, make timely and effective decisions, and take purposeful action. This cyclic process is known as the OODA loop: Observe, Orient, Decide, and Act.[3]

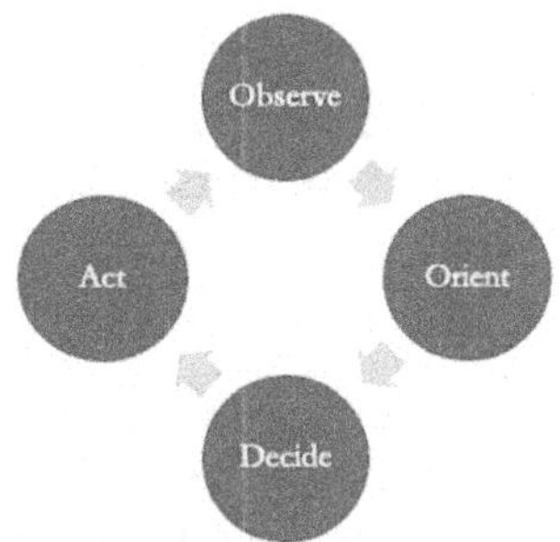

Military decision making — the OODA loop.

The side that can cycle through the OODA loop more efficiently and rapidly gains the advantage, forcing its opponent to react under pressure and on unfavorable terms. Swift, decisive action, even if imperfect, can disrupt and destabilize an opponent, further degrade their decision-making, and prevent them from setting and pursuing their own objectives. In many cases, this agility matters more than perfection. Taking timely action, even if flawed, can yield a crucial advantage.

Corollary of Maxim 11: A mediocre plan executed violently is better than a brilliant plan executed late.[4]

By consistently outthinking and outmaneuvering the adversary, a force can seize the upper hand, leaving its enemy off balance and unable to adapt to a developing situation. This enables the more agile force to retain the initiative, forcing the opponent to react and preventing them from advancing their own objectives. In this way, a well-informed and flexible force may overcome a formidable yet rigid adversary.

However, each decision is a hypothesis; a proposed solution to the current situation based on imperfect information. Therefore, speed alone is not enough; decisions must also be effective. Success depends on a combatant's ability to remain synchronized with the evolving situation, the environment, and the enemy's disposition.

For example, launching a swift attack may catch an opponent by surprise. However, if the enemy is gathering to refuel and rearm, waiting could offer a greater advantage by simultaneously striking when they are most vulnerable and destroying their critical logistics. Attacking a disorganized force can catch them off guard, reduce their resistance, instill fear, and erode their confidence and morale. While speed is crucial, timing and rhythm may be just as vital.[5]

Maxim 21: Strike when and where an opponent is vulnerable. Exploit opponents when they are unprepared, confused, or in disarray.

Humans are efficient at managing small groups but often struggle with more than five subordinates. To overcome this limitation, military organizations usually *command* and *control* their forces using a hierarchy in which each individual is responsible to their immediate commander. Such a structure enables quick, efficient command and control of large organizations.

A commander may organize a military force hierarchically based on its abilities and mission. The smallest ground unit is a fireteam (or section), composed of two to four soldiers led by a corporal. A squad consists of two or more fireteams led by a squad leader, typically a non-commissioned officer. Above squads are platoons, usually led by a lieutenant, composed of three or more squads and sometimes heavy weapons. Next are companies (or batteries for heavy weapons units) led by a captain, containing three or four platoons with additional heavy weapons. Larger formations are battalions, usually commanded by a lieutenant colonel. If the unit requires external support, it may be designated a brigade, led by a colonel or brigadier general. A self-sufficient unit of similar size is often called a regiment. Above brigades and regiments are legions or divisions, followed by armies, each progressively larger and more strategically significant.

Aviation units may be organized into sections, flights, squadrons, groups, wings, and air forces.

Naval units are often organized around their vessels and the missions they are tasked with performing. They may be organized into divisions composed of two or more ships, then squadrons, groups, and fleets.

Leaders may adapt military structures into special formations based on mission requirements. A ground force may be reinforced with engineers, artillery, heavy armaments, armored forces, and transportation, forming a *task force* or *strike group* tailored for a specific operation.

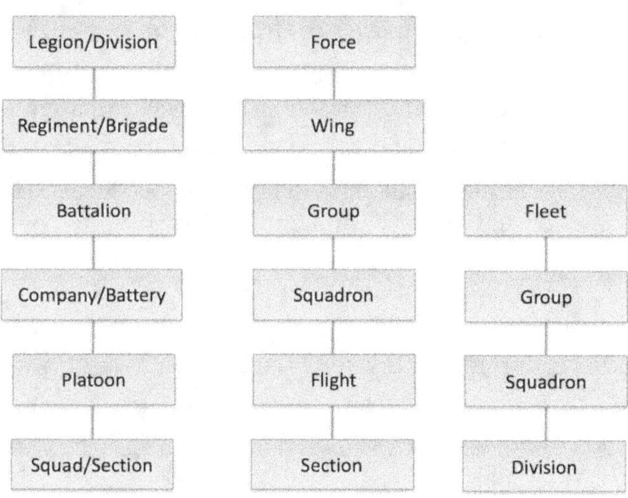

Typical ground, air, and naval force hierarchies.

A military leader is often portrayed as a stoic figure who calmly assesses the situation and issues commands that subordinates dutifully obey. However, a hierarchy is not always required. An organization can operate through a heterarchy, a structure in which authority and decision-making are shared across multiple individuals or units rather than concentrated in a single chain of command. In a heterarchy, roles and influence may shift depending on the situation, allowing for flexibility and collaboration, provided the group can still make timely and effective decisions in combat.

Subversive movements frequently adopt decentralized cells that possess limited knowledge of one another. This arrangement reduces the impact of compromise should individual cells be exposed, captured, or destroyed.

Similarly, leadership does not have to be autocratic. A leader's style depends on the individual, their subordinates, and the norms of their society. What matters most is the ability to make rapid, sound decisions and act decisively, regardless of the structure or style employed.

The figure below shows a continuum of control within an organization. On the left side, minimal control implies greater decentralization, allowing subordinate units more autonomy, flexibility, and initiative; however, it also risks fragmentation, inconsistent execution, and loss of cohesion. On the right, increased control reflects centralized decision-making, tighter coordination, and greater predictability, but it can lead to rigidity and suppressed initiative. This spectrum highlights how different leadership styles and command structures vary in the degree of oversight they apply, suggesting that the appropriate level of control depends on the mission, environment, and capabilities of the forces involved.

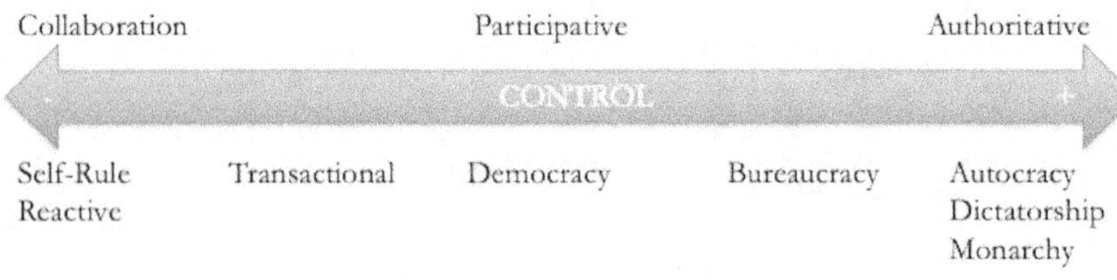

The spectrum of control.

The ant-like alien Formics of **Ender's Game** communicate via their hive mind but exhibit limited independent thought and action. They operate under an absolute autocracy in which their Hive Queen centrally controls all the drones and provides unity of command.

Opposing the Formics, Ender organizes his jeesh, an Arabic word meaning 'army,' into a decentralized and highly adaptive heterarchy of toons, which are smaller platoon-sized units. He forms these toons based on mission requirements and individual abilities. This structure allows toon leaders to exercise initiative in organizing, training, and executing their assignments, while Ender retains overall command.

While maintaining strategic oversight, Ender grants his subordinates the freedom to adapt to battlefield conditions, encouraging creativity and independent action. He is able to lead this way because his subordinates are hand-picked, highly trained, and represent the top individuals from their society. His ability to match leaders to roles effectively, combined with their adaptability, enables him to shift the battle dynamics and constantly outmaneuver the Hive Queens (Card 1999).

One of the primary challenges facing human organizations lies not in the formal command arrangement but in human nature. Personal ambitions, fears, egos, and insecurities can disrupt or overshadow an organization's objectives, posing a direct threat to its mission.

In the prequels to *Ender's Game*, humanity confronts the first Formic invasion of the solar system. This is a fight for survival complicated by internal dysfunction. Poor leadership and bureaucratic inefficiencies, both within and beyond the government, undermine the military's effectiveness. Although military leaders remain outwardly loyal, their decisions are often shaped by personal ambition and self-interest rather than by military success, with disastrous consequences.

Command becomes complicated when subordinates report to multiple leaders, especially when groups with differing objectives must cooperate. This often arises in alliances where members pursue divergent objectives or come from distinct cultural backgrounds, often leading to friction and fragmented military operations.

In *Star Trek: Voyager*, a group of Federation colonists form the Maquis Resistance after being forcibly displaced by a peace treaty between the Federation and Cardassia. Maquis raids inflame tensions with the Cardassians and risk dragging the Federation back into war. Starfleet dispatches the *USS Voyager* to apprehend the Maquis raider *Val Jean*. When the *Voyager* discovers the *Val Jean*, a powerful alien known as the Caretaker, accidentally, but violently, transports both ships 75,000 light-years across the galaxy. The event kills the *Voyager*'s first officer, flight controller, chief medical officer, and chief engineer. The Maquis leader, Chakotay, sacrifices his vessel to save the *Voyager*. Crippled and far from home, the former adversaries combine their crews aboard the *Voyager*, establish a mutually acceptable command structure, and work together to find a way home (Berman, Piller, and Taylor 1995).

When a member of the Marquis crew is suspected of betraying the Federation by sharing classified information with an extraterrestrial civilization, Chakotay, alongside his former comrades, aids the Voyager's security chief in the investigation. Despite initial tensions stemming from the merger of the two crews, Chakotay puts their differences aside and undertakes to unify the Voyager's crew under a single command (Coyle 1995).

This exemplifies that a leader may need to navigate, negotiate, or eliminate internal conflicts that obstruct effective operations, including political rivalries that push competing agendas, economic pursuits that weaken military readiness, and bureaucracies and service rivalries that compete for 'their share' of funding and operations, to the detriment of the strategic ends.

In rare cases, infighting and outright insubordination may result in more effective operations. In the early stages of *Ender's Game*, the authoritarian Rat Army commander, Rose the Nose, resented having Ender in his unit. He was dismissive, openly hostile, and belittled him in front of others, even restricting his training. However, his subordinate, Dink Meeker, broke ranks and encouraged Ender to continue his studies and training, directly disobeying Rose's orders. Although Dink's actions were insubordinate, he acted to strengthen his troops. Though unconventional, his defiance illustrates how challenging authority in pursuit of effectiveness can sometimes lead to unexpected success.

However, insubordination rarely ends well. Captain (and later Admiral) James Kirk of the *Star Trek* universe is ostensibly an explorer; however, his military prowess is unmatched. This is evident in numerous situations, including the prolonged cat-and-mouse pursuit of a stealth-enabled Romulan Bird-of-Prey terrorizing the Neutral Zone (McEveety and Schneider 1966). Kirk becomes famous for his innovative tactics, even having starship maneuvers named after him, such as the "Pattern Kirk Epsilon" evasive maneuver (Abrams 2002).

Kirk is an expert in the use of deception to avoid combat or distract an opponent until he can gain an advantage. In two incidents, he bluffs his way out of impossible situations using a fictitious self-destruct weapon (Pevney, Roddenberry, and Harmon 1967; Sargent and Sohl 1966). His most elaborate and audacious misdirection occurs when he is ordered to steal a Romulan cloaking device. He feigns mental instability to mislead his own crew, relies on his first officer's staged betrayal to distract the Romulans, and uses a surgical disguise to infiltrate their ship. Each layer of misdirection conceals his true objective until it is too late to stop him, making this his most sophisticated use of deception. (Fontana 1968).

Kirk is renowned for his adventurous spirit and unwavering optimism. He is not afraid to challenge the status quo or question authority, and he consistently seeks innovative solutions to complex problems. His belief in others' potential is a driving force behind his leadership and decision-making.

Kirk's rebellious optimism is clearest in his refusal to accept a no-win scenario. He treats every crisis as a puzzle with a hidden solution, defying the limits imposed by circumstance or authority. For him, survival and victory are matters of ingenuity and willpower, not inevitability. This is most evident in his reprogramming of the Kobayashi Maru test, a seemingly unwinnable simulation designed to evaluate Starfleet Academy cadets' character by placing them in an unwinnable situation. During the test, cadets encounter a critically damaged civilian fuel ship, the Kobayashi Maru, stranded in the volatile Neutral Zone between the Federation and the Klingon Empire. Kirk faces an impossible choice: attempt a dangerous rescue and jeopardize his ship and crew, or abandon the Kobayashi Maru to its fate. When Kirk faces this test, he refuses to accept the outcome. He rewrites the simulation to make it winnable, demonstrating his belief in overcoming insurmountable odds. This act not only challenges the established Starfleet training protocol and ethics code but also reveals Kirk's unwavering spirit and uncompromising optimism. Spock, acknowledging the novelty of Kirk's solution, remarks, "It was a unique solution." Kirk simply replies, "It was unique because it had never been tried" (Abrams 2013).

There is no better compliment to a leader's military prowess than to be recognized by their opponents. Klingon General Chang reveled in his attack on Kirk at the Battle of Khitomer until the tides of war turned against him (Meyer et al. 1991). The legendary Klingon military leader Kor surmises that a battle against Kirk "would have been glorious" (Newland, Coon, and Roddenberry 1967). Likewise, Captain Klaa believes defeating Kirk will make him the greatest warrior in the galaxy (Shatner et al. 1989). Lieutenant Commander Jadzia Dax also comments that the legendary Klingon military strategist Koloth always regretted not having the chance to face Kirk in battle.

Kirk understands the ways and means required to achieve his objectives and ends, but his methods often conflict with Starfleet's policies and doctrine. In *Star Trek: Into Darkness,* a young Kirk learns a hard lesson when his actions lead the primitive natives to observe his starship, violating Starfleet's Prime Directive, which prohibits interference in other societies. Later, he risks his crew and ship and attempts to cover up his mistakes by omitting key information from his report to Starfleet. Admiral Pike admonishes him and relieves him of his command, stating, "You don't comply with the rules, you don't take responsibility for anything, and you don't respect the chair" (Abrams 2013). Pike is referring to a leader's need to answer to a higher calling: respect for authority, the responsibilities entrusted to them, and the lives of the personnel under their command. This is an early lesson for Kirk, who learns to be responsible and selfless.

Corollary of Maxim 4: The military is trusted with defending the sanctity and security of society. The mission must come first.

INITIATIVE

"Doing nothing is just as bad as doing the wrong thing."

— ALEX KAMAL, PILOT, ROCINANTE (EISNER ET AL. 2017)

In the *Terminator* universe, humans develop an advanced computer system called Genisys, which becomes self-aware and evolves into Skynet. Skynet determines humanity is a threat to its existence. On

'Judgement Day,' it seizes control of the global computerized military systems and launches a surprise nuclear exchange, eliminating the world's military and civilian population centers. It follows this with mop-up operations using its robotic armies.

In response, humans establish a resistance movement led by John Connor, waging a war against Skynet's automated forces. When the humans breach Skynet's defenses and threaten its existence, Skynet responds by triggering its doomsday plan, activating the Time Displacement Equipment to send a T-800 android Terminator disguised as a human back in time. Its mission is to eliminate Sarah Connor, the mother of the future leader of the human movement, thus eradicating the resistance before it can begin.

However, Skynet makes a crucial mistake. After sending the Terminator back in time, it fails to destroy the time machine. Human forces capture the facility and use the device to send their own operative, Kyle Reese, to protect Sarah Connor and stop the Terminator. The T-800 relentlessly pursues Sarah, but in doing so, inadvertently sparks her relationship with Kyle Reese, leading to her pregnancy with John Connor, the very leader destined to destroy Skynet. In a twist of fate, Skynet's attempt to erase its nemesis results in its creation (Cameron 1984).

In *Terminator 2: Judgement Day*, Sarah and John are targeted by a T-1000 mimetic polyalloy, a liquid metal shapeshifter programmed to kill them. In response, the future John Connor sends back a reprogrammed T-800 to protect them. Realizing that they are merely reacting to a nearly unstoppable foe, Sarah and John change their approach. Instead of waiting to be hunted, they seize the initiative and launch a mission to eliminate the engineer responsible for Skynet, aiming to prevent its creation and avert the future apocalypse (Cameron and Wisher 1991).

By applying Maxim 13, seizing and maintaining the *initiative*, a leader shapes the conflict by choosing time, place, and manner of action, pursuing their own objectives rather than reacting to conditions shaped by the enemy. Initiative allows a leader to control the battlefield and impose their strengths against an opponent's weaknesses. Doing so requires a clear understanding of one's capabilities and limitations, insight into the adversary's capabilities and intentions, foresight to anticipate developments, and the ability to act swiftly and decisively.

Scouts, spies, and intelligence operations help reveal enemy plans and identify the optimal time and place to strike. Once control is gained, they may sustain the operational tempo, or OPTEMPO, to build momentum. Maintaining pressure on an opponent can restrict their options, reduce their resources, and keep them off balance and reactive. The alternative is to remain passive and cede control of the conflict to the opponent.

Corollary of Maxim 13: Seize the initiative through insight, foresight, and decisive action.

Perspective-taking is a technique that focuses on understanding an opponent's mindset, motivations, objectives, and likely actions to anticipate and influence their behavior, thereby gaining a strategic advantage. It does not require sympathizing with or approving of an adversary but emphasizes comprehending their perspective and decision-making. This approach involves cultural, ideological, and historical study; psychological profiling of enemy leaders; and strategic and tactical analysis of potential plans, actions, and deception efforts, all aimed at identifying, evaluating, and ultimately defeating the opponent.

Grasping a human opponent's mind is challenging; deciphering an alien one, shaped by entirely different biology, culture, history, and technology, may produce a worldview beyond human comprehension. Empathizing with their expressions, especially when those differ from our own, demands emotional sensitivity and openness. Our biases and assumptions further hinder the capacity to comprehend and adopt an alien perspective.[6] Successfully navigating this intellectual, emotional, and ethical minefield requires a rare combination of intelligence, creativity, and unwavering mental fortitude.

A leader can seize the initiative by acting first, forcing the opponent to react, and striking again to maintain control and steer events toward a desired outcome. Offensive action is the most effective and decisive method for a military force to gain and hold the initiative while preserving freedom of action. In contrast, a defensive position allows the attacker to dictate the time, place, and conditions for a fight.

A skilled commander seizes the initiative by remaining agile and by recognizing and exploiting opportunities as they arise. As Maxim 21 notes, striking when an opponent is vulnerable increases the likelihood of achieving a decisive victory. Adversaries often expose themselves as they adapt to new objectives, recover from setbacks, or replace fallen leaders, creating brief openings that are ideal for exploitation. Commanders who strike during transitions or movements, such as when enemy forces are repositioning, assembling, or refueling, can exploit these moments of instability.

Bold, deliberate maneuvers across the battlefield can disrupt an opponent's plans by imposing unexpected pressure and forcing a rapid response, often without sufficient preparation. By advancing with purpose and exploiting gaps in formation or timing, a commander can create confusion, strain command structures, and provoke hasty decisions that erode cohesion and effectiveness.

Maintaining a reserve force preserves flexibility. If the main body becomes decisively engaged, the uncommitted reserve can exploit vulnerabilities, reinforce success, or recover from setbacks.

Maxim 22: Reserves provide battlefield flexibility.
A reserve force allows commanders to adapt quickly and respond effectively to changing conditions.

Alternatively, a leader may surprise an opponent using stealth and deception. Unexpected actions can thwart an opponent's plans and introduce confusion, fear, and panic. These techniques are explored further in Chapter Thirteen.

A common myth holds that a leader must seize the initiative. In reality, a leader may hand the initiative to an opponent, whether intentionally or by mistake, through poor decisions, inaction, or external constraints. For example, failing to conduct adequate reconnaissance or gather intelligence may lead a force into a decisive engagement on the enemy's terms. Political pressure, alliance commitments, or logistical constraints may compel a leader to react instead of act, placing the opponent in control of the tempo and conditions of the conflict.

Consider a commander who positions their forces in a defensive location with no means of escape. This decision restricts their future actions, leaving them to deter an attack, hold their ground, surrender, or fight to the death. By committing to a static defense without flexibility, they forfeit the ability to maneuver or withdraw when conditions are more favorable. This violates the intent of Maxim 10, which stresses preserving future options. A leader should avoid choices that limit flexibility and shift control of the engagement to the adversary.

Corollary of Maxim 10: Choices that restrict future options may reduce a force's initiative.

Thus, a leader gains and retains the initiative by rapidly exploiting emerging opportunities, maneuvering faster than the opponent can effectively respond, not allowing the main body to become committed before its time, maintaining a reserve force, and using intelligence to clarify an opponent's location, force, and intentions. These actions enable a leader to anticipate an opponent and rapidly respond to changing conditions.

In *Star Wars: Episode VI – Return of the Jedi*, during the Battle of Endor, Lando Calrissian realizes that the second Death Star is already fully operational when the Rebel fleet comes under immediate, accurate fire. This revelation invalidates the Rebels' assumption that the space battle would merely delay the Imperial response while the ground team disabled the planet-based shield generator. Recognizing that the ground assault required more time than anticipated, Lando makes the critical decision to shift the Rebel fleet's efforts to abandon a passive *holding action* and instead commit the Rebel fleet to close engagement with Imperial capital ships.

This transformed the space battle from a delaying action into a decisive fight for survival. By engaging the Imperial fleet directly,

Lando prevented the Death Star from systematically destroying Rebel ships while buying the ground team the time it needed to succeed. The decision also forced the Imperials into a dense, chaotic battlespace that blunts their numerical and firepower advantages. Lando's adjustment illustrates effective battlefield adaptation in the face of intelligence failure, demonstrating how timely decision-making and tactical flexibility can preserve, or even seize, the initiative after initial assumptions collapse.

Guerrillas and rebels often rely on maneuver and surprise, employing small hit-and-run operations against traditional military forces. They may disperse and hide until they can mass in a surprise strike against a vulnerable opponent. Then, they must vanish before the opponent can respond and destroy them with their overwhelming mass. Their larger opponents may believe they maintain the initiative with their lumbering forces able to go where they wish, but the guerrillas also have the initiative to surprise in small-scale strikes.

Maxim 23: A battle is often decided before the first round is fired. Preparation and planning are critical to success.[7]

While maintaining the initiative often provides a significant advantage, it can become a liability. Consider the following points:

- Reliable Strikes Are Essential: Offensive action succeeds only if weapons consistently strike and damage intended targets; attacking an enemy able to evade or absorb damage risks defeat.
- Actions Must Support Strategy: Offensive actions should align with broader objectives to avoid unnecessary losses, conserve resources, and prevent long-term setbacks.
- Surprise Reduces Predictability: Predictable moves invite countermeasures, while surprise increases effectiveness and limits enemy preparation.

- Avoid Overextension: Sustained operations strain forces and can lead to a *culmination point* where they become exhausted and vulnerable.
- Intelligence Guides Decisions: Incomplete intelligence increases risk, while accurate, timely information prevents costly errors and ambushes.
- Respect the Enemy: Underestimating an opponent's capabilities or resolve can turn initiative into vulnerability.
- Weigh the Consequences: Aggressive offensive operations can provoke political or strategic backlash that outweighs tactical success, escalating a conflict, drawing in new adversaries, or triggering economic retaliation that harms the aggressor's home front.

Initiative can be a double-edged sword. It is most effective when combined with careful planning, surprise, adaptability, and an understanding of the overall environment. Missteps can turn the initiative into a liability, leaving it vulnerable to exploitation by a prepared opponent.

This chapter examined the complexities of military command and the principles that underpin its effectiveness. These principles focus on unified means, an efficient command structure, and actions that enable a force to seize and maintain the initiative. However, the effectiveness of military command depends on multiple factors, including political and social structures, available resources, and the nature of opposing forces. The next chapter explores the mechanisms that forge military power itself.

CHAPTER 9

MILITARY POWER

"The power to cause pain is the only power that matters, the power to kill and destroy, because if you can't kill then you are always subject to those who can, and nothing and no one will ever save you."

— ENDER WIGGIN, STUDENT, BATTLE SCHOOL

(CARD 1999)

War is the pursuit of strategic objectives and ends through military means, seeking to achieve the greatest impact on the enemy with minimal expenditure of resources and lives. This chapter explores how societies concentrate and preserve military power.

MASSING MILITARY POWER

Victory depends on the ability to synchronize the effects of military power at a decisive place and time to overwhelm and dismantle an opponent's forces. Success lies not only in possessing strength, but in the precise concentration and application of that strength when and where it matters most. A military force builds toward this by cultivating discipline, conducting rigorous training, executing well-planned maneuvers, and employing deception to mislead and disorient the enemy. These elements allow a force to shape the battlefield and bring power to bear under favorable conditions.[1]

At the same time, preventing the enemy from exploiting opportunities is equally vital. Disrupting their communications, delaying or confusing their movements, undermining their morale, and interfering with their logistics can deny them the ability to coordinate and concentrate their forces. By preventing an enemy from concentrating forces, a commander limits the opponent's freedom of action and reduces the enemy's capacity to strike effectively. Mastery in war requires maximizing one's own combat potential while systematically degrading the adversary's.

Corollary of Maxim 11: Synchronize the effects of military power at a decisive place and time to overwhelm an opponent's forces.

It is important to distinguish between massing *military power* and massing *forces*. A leader's goal is not to bring all their military units and logistics to the same place, but to synchronize their effects at the decisive place and time. Imagine an interstellar clash: a fleet commander might launch slower missiles, followed by hypervelocity rounds, and finally light-speed beam weapons. This onslaught, if meticulously timed, may simultaneously overwhelm the enemy's defenses in a single, unstoppable wave, rather than arrive piecemeal and allow their opponent to absorb, dodge, or deflect them.

Likewise, *firepower* need not be lethal to be effective. The value of firepower lies in its ability to influence an opponent's behavior, not just in its capacity to destroy. A single stormtrooper with a blaster set to stun may possess enough firepower to subdue an entire Jawa clan, assuming the trooper can hit its targets. Here, the utility of firepower lies in its ability to neutralize resistance without causing permanent harm.

The presence or threat of firepower alone may achieve military objectives. A well-timed display of force, such as armored vehicles or combat aircraft, or the audible charging of weapons, may cause an opponent to retreat, surrender, or fall into disarray. Firepower includes both its destructive capability and its psychological impact. Used wisely, it can compel an adversary to act against their own interests without the need for direct engagement.

One technique for concentrating firepower is to position forces in mutually supporting locations with *overlapping fields of fire* and *overwatch*. Overlapping, or interlocking, fields of fire occur when two or more weapons can engage the same target. The area these combined weapons cover is called the *beaten zone.*

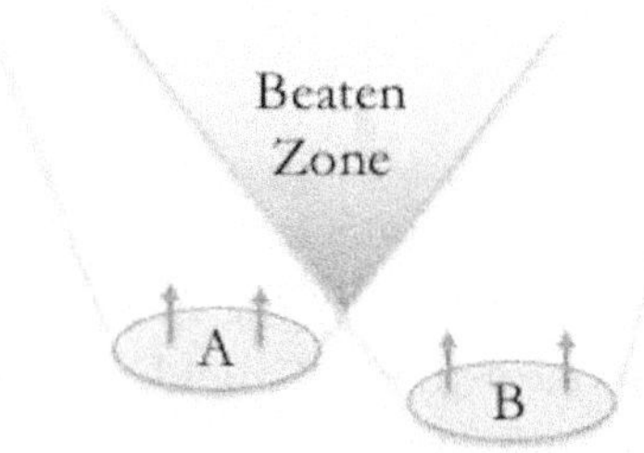

Teams A and B weapons creating a beaten zone.

An overwatch occurs when a unit maneuvers while another provides security and protective fires. In a bounding overwatch, some units remain stationary to engage targets while others advance, shift to a flank, or fall back. Once the moving elements reach a new position, they halt and provide covering fire as the previously stationary units maneuver. This leapfrog approach offers two key advantages: it prevents the entire force from becoming decisively engaged and ensures continuous enemy suppression while other elements reposition.

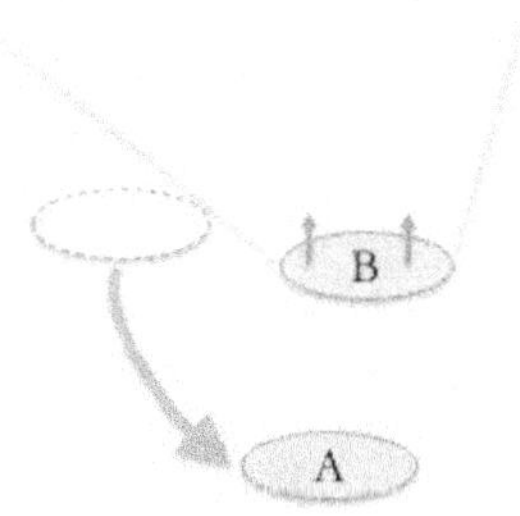

Team B provides overwatch while Team A moves to the rear.

To mass firepower, a leader may concentrate military forces in proximity to one another. However, as additional units are concentrated in a given area, they may begin to interfere with one another's targeting and maneuvering, and may overwhelm local facilities and resources.

Leaders must also consider the threat posed by an adversary's standoff weapons and weapons of mass destruction, which could target concentrated forces. Facing such destructive capabilities may require dispersing forces across a battlespace and massing them only when necessary to achieve overwhelming force at a decisive moment.

Massing combat power can create large-scale, intense violence that shocks and demoralizes an opponent, overwhelming their senses, distorting their perception of the battlefield, and eroding their will to resist. When fear takes hold of a single soldier, it can quickly spread throughout the ranks, transforming individual panic into collective terror.

However, the purpose of this shock is not intimidation for its own sake. It is a tactical method for setting conditions, whether by enabling a follow-on attack that exploits the enemy's disarray or by convincing them that defeat is inevitable, compelling them to flee, surrender, or negotiate.

Maxim 24: Unchecked fear breeds ruin.
Fear can paralyze individual soldiers, then spread as panic until it
consumes an entire unit.

Disrupting and demoralizing an opponent's military, population, and leadership can be accomplished in several ways. One method is to separate and isolate enemy forces, creating the perception that their defeat is inevitable. Another is to threaten retreat routes, inducing panic by endangering their escape, resupply, or reinforcement. Campaigns may also target what a society holds sacred, such as its capital, religious sites, major hospitals, or homeworld, in order to provoke fear or despair. Finally, interfering with the production or distribution of essential resources, including food, fuel, or medicine, can steadily erode combat effectiveness and civilian morale, reducing the will to resist.

A real or fictional history of punishing those who resist may instill terror among adversaries. A frightened and demoralized populace may implore leaders to end hostilities, disrupt transportation while fleeing, and burden support structures. In **Star Trek: Deep Space Nine**, the Jem'Hadar's ruthless reputation compels societies to surrender and submit to their rule. Accounts of brutality, atrocities, and swift retaliation serve as powerful deterrents, steadily eroding the will to resist invasion.

However, the use of terror can backfire. If warriors believe their defeat is inevitable, histories of atrocities may convince them to fight to the last. In the **Star Wars** universe, Emperor Palpatine believes terror will prevent the planets and systems from uniting into a large-scale rebellion. The destruction of the planet Alderaan, along with most of its three billion inhabitants, no doubt caused intense fear among the Empire's surviving citizens. Yet, in demonstrating his power, Palpatine also revealed the dangers of an unchecked empire, inspiring the Rebels and much of the galaxy to resist him and his followers.

One method to mass military power is to exploit a *comparative advantage* between friendly and enemy forces. A comparative advantage occurs when one group can create effects that another group cannot reproduce or effectively counter. Imagine a force using effective, coordinated long-range standoff weapons, such as snipers and artillery, to strike targets within its kill zones, leaving the enemy powerless to retaliate. This one-sided engagement whittles down the opponent, instills intense fear, and undermines morale.

Warriors threatened by standoff weapons may be forced to advance while under fire, using available cover and concealment, until they can engage or suppress the enemy's weapons or escape the area. The mere threat of such weapons can deter forces from entering potential kill zones, prompting them to disperse to minimize casualties, which may limit their ability to mass fires effectively. The presence of standoff weapons may deter intrusions or compel adversaries to avoid the area entirely. A cunning warrior may exploit this by guiding opponents toward zones that appear unprotected, only to destroy them once they enter the hidden kill zones, or by tricking them into avoiding areas they wrongly believe are dangerous.

Sometimes, standoff weapon operators cannot directly observe their targets and must rely on reconnaissance units, guerrillas, patrols, drones, or remote sensors to locate, track, and report them. Just as soldiers depend on food, vehicles require fuel, and political leaders rely on public support, this dependency between spotter and shooter creates a vulnerability. Finding and neutralizing observers undermines the effectiveness of standoff weapons without requiring direct confrontation with these weapons.

Corollary of Maxim 6: Destroying or disrupting a critical resource or support system weakens all forces dependent upon it, creating cascading effects throughout their operations.

It is crucial that military strategists identify similar dependencies between supported and supporting forces. Consider an invading fleet supporting a ground force: the fleet might support an assaulting ground force, but the ground force cannot support the fleet. Attacking the fleet might cripple both the fleet and its ground support, simplifying the defeat of the latter. Conversely, targeting the ground force leaves the fleet untouched. Therefore, striking the fleet yields a greater immediate effect.

In *The Expanse* universe, the Earth leadership understands the power and threat that the Martian Marines pose to Earth. United Nations Undersecretary General Avasarala notes that despite the Martian lighter gravity, the Martian Marines "always train in one G, Earth gravity…it won't end out there. If we continue to force their hand, sooner or later, they will invade." Undersecretary Errinwright counters, "Mars is an island. If we cripple their navy…[the Marines] can't touch us" (Franck, Abraham, and Fergus 2017). Errinwright understands that eliminating the Martian fleet removes the dependent Martian Marines from the equation.

A commander seldom has the opportunity to defeat an enemy's entire force simultaneously. Even if such an engagement occurs, the opponent might still prevail through superior military power, advantageous terrain, stronger discipline or leadership, or simply fortuitous circumstances. A commander can, however, achieve *local combat superiority* by concentrating an overwhelming force at a single decisive point. Successfully massing forces against an opponent requires maneuvering swiftly to strike and defeat that enemy faster than the opponent can observe, orient, decide, and act. After defeating their immediate opponent, the victorious force can then shift rapidly to the next target. This approach, known as *divide-and-conquer* or *defeat in detail*, allows a weaker force to wear down a much stronger opponent by engaging and destroying isolated portions of the enemy incrementally.

The divide-and-conquer strategy has three forms. In the first, a commander splits the enemy into separate pieces and then masses military power to defeat each isolated piece independently. This approach allows a smaller, cohesive force to defeat a numerically superior but divided opponent, or for a larger force to defeat groups of smaller forces while minimizing its losses.

Captain "Black Jack" Geary demonstrates the divide-and-conquer strategy during his running battles with the Syndicate fleets in *The Lost Fleet* series. Outnumbered and commanding a battered force, Geary uses precise maneuvering to isolate portions of his pursuer's fleets. By concentrating his firepower on a single section of the Syndicate formation, he overwhelms and destroys those ships before the rest can respond effectively. This tactic prevents the enemy from leveraging their superior numbers, turning their poor coordination into a decisive advantage for his own outmatched fleet.

The second form also involves dividing the enemy, but emphasizes neutralizing parts of the opponent's forces without resorting to direct combat. Commanders achieve this by exploiting fear, promoting apathy, or demoralizing enemy ranks until they disengage voluntarily. A leader may then concentrate resources to eliminate the remaining active threats.

Geary relied on deception and guile to outmaneuver the Syndicate leadership as he led his fleet back toward Alliance space. By concealing his fleet's route and feigning movements toward multiple destinations, he forced the Syndics to divide their forces to cover every avenue of pursuit. This scattering of strength left individual Syndicate task forces vulnerable, allowing Geary to engage and defeat smaller, isolated groups rather than facing the full weight of Syndicate power at once. Through calculated misdirection, he turned enemy uncertainty into a weapon, ensuring that each clash favored his battered but disciplined fleet.

The last form, sometimes called a bait-and-bleed strategy, involves dividing a larger enemy force into multiple factions and inciting them to turn on each other by sowing distrust or exploiting their differences. This approach compels adversaries or allies to weaken themselves through internal conflict, enabling a commander to defeat the diminished forces separately and decisively.

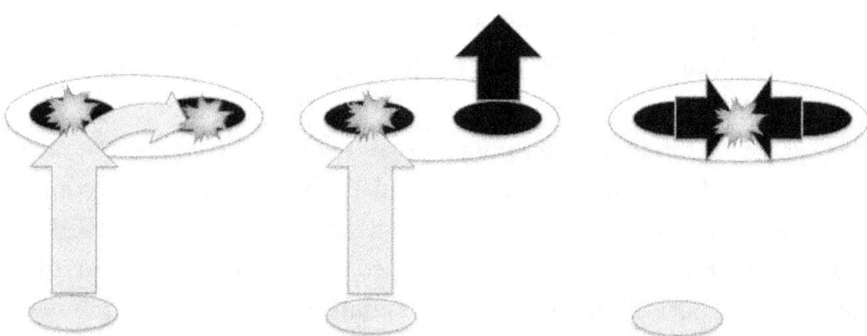

First, second, and third forms of divide-and-conquer (from left to right).

In ***The Lost Fleet,*** the Enigma aliens secretly provoke the two rival human factions, the Alliance and the Syndicate Worlds, into a prolonged war (Campbell 2011). Despite possessing vastly superior propulsion and communications technologies, the aliens avoid direct confrontations. Instead, they employ deception and betrayal to entice the two into a century-long genocidal war. As a result of the bait-and-bleed strategy, the aliens encourage humanity to exhaust its limited resources in continuous warfare rather than investing in growth and expansion.

A divide-and-conquer strategy enables a force to mass strength against isolated segments of an opponent, while preventing that opponent from consolidating its power, but a strategist must remain vigilant to ensure their own forces do not become vulnerable to division and piecemeal defeat.

Corollary of Maxim 21: Mass on the weak and destroy them. Divide opponents and conquer the smaller pieces. Complement of Maxim 21: Concentrate forces or be destroyed piecemeal.

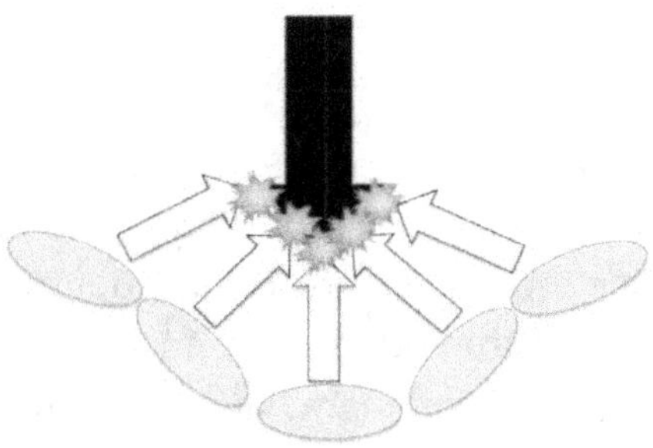

Defenders (bottom) massing on an attacker (top), as seen from above.

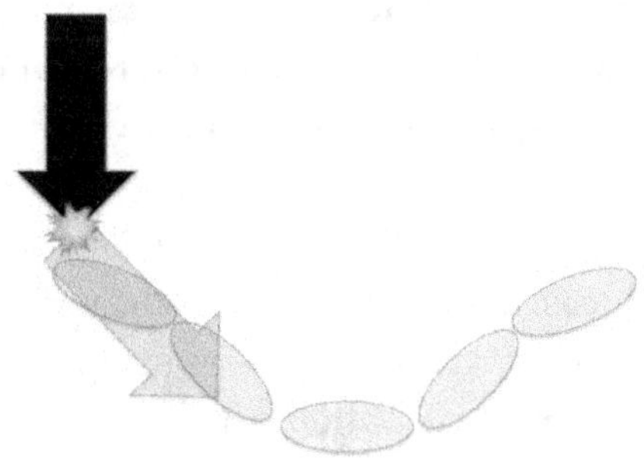

An attacker (top left) using divide-and-conquer to defeat a larger defender, as seen from above.

A leader may employ various methods to mass forces effectively. A *unified assault* brings units together into a cohesive force to attack an opponent. A *convergent assault* coordinates multiple forces to strike an opponent simultaneously from multiple directions. Finally, a swarm is an *omni-directional assault* involving many smaller units attacking simultaneously from multiple directions.

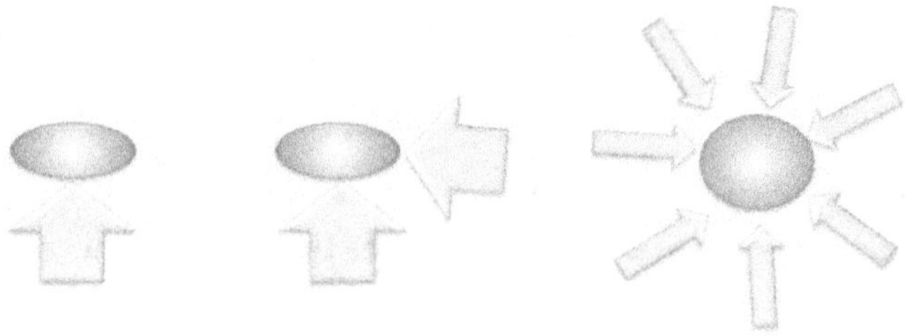

A unified, convergent, and omni-directional assault (left to right).

Having the largest and most powerful military goes a long way towards victory, but does not guarantee success. Likewise, possessing a smaller or weaker force does not ensure defeat. An inferior force should avoid direct confrontation with an opponent's strength and instead seek means and ways that change the situation and shift the advantage in its favor.

Maxim 25: When faced with a no-win situation, change the situation. Rather than accepting defeat, take action to reshape the conditions and create opportunities for victory.

There are many ways to entice a stronger opponent into attacking on terrain that reduces their advantages. A defender might position troops behind the crest of a hill in a reverse-slope defense. This forces attackers to expose themselves as they advance over the top, nullifying their long-range weapon advantages and allowing the defender to mass fire at close range.

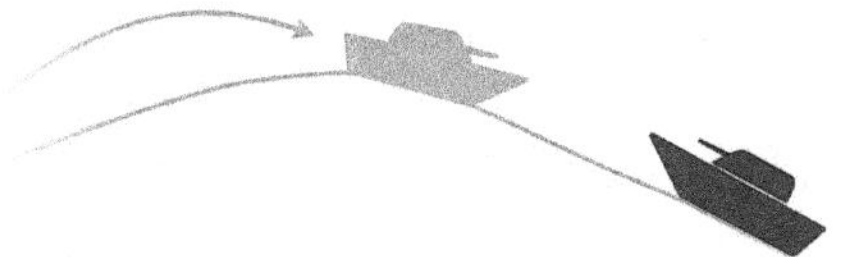

A reverse slope defense (right) shields the defender from long-range fires until the attacker (left) enters short-range weapon range, observed from the side.

Similarly, defending behind a choke point compels an attacker to funnel troops through a restricted area, allowing the defender to mass firepower effectively against a narrow front. Difficult terrain, such as mountains or swamps, reduces mobility, while urban environments reduce mobility and weapon effectiveness, forcing units to congregate and become vulnerable to area weapons.

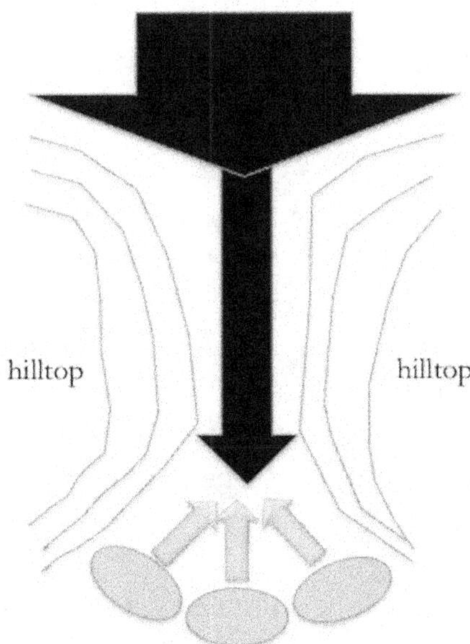

A choke point restricts the number of attackers (top) that can attack simultaneously, as seen from above.

Numerically weaker forces can avoid destruction by dispersing to evade detection, then rapidly concentrating to deliver swift, unexpected strikes. This approach is especially effective in subversive warfare, such as guerrilla hit-and-run tactics. Alternatively, a smaller force may employ superior firepower or weapons of mass destruction to inflict *asymmetric effects*.

To offset weakness, a leader should align with groups that share similar objectives, increasing their combined military power and sharing both the costs and potential gains of a conflict. Conversely, weakening an adversary's alliances before launching military action can significantly reduce their strength.

ECONOMY OF FORCE

"The queen doesn't sacrifice everything for a pawn," Sam observed during a Spartan training exercise, hinting at the cold logic of command. John answered, "All pieces return to the same box in the end."

— JOHN (117) AND SAM (034), SPARTANS (NYLUND 2001)

A military rarely has enough troops, weapons, equipment, ammunition, supplies, or time. Adversaries, the environment, and misfortune further conspire to diminish resources. Conserving military power preserves flexibility, resources, and strategic choices, ensuring a force can respond effectively to future uncertainties or opportunities. Leaders must therefore safeguard their means and apply power in the most effective and efficient ways possible to achieve objectives while avoiding unnecessary expenditure.

Corollary of Maxim 10: The goal of war is to achieve objectives while expending the least possible means.
Complement of Maxim 10: Every unit of power consumed reduces resources for future military, economic, and political objectives.

Force multipliers, sometimes called combat multipliers, are tangible and intangible means and ways that amplify or reduce military power disproportionately relative to the effort applied. Tangible force multipliers include special military units. Engineers, also known as pioneers or sappers, facilitate friendly unit movements, hinder enemy maneuvers, and construct and destroy structures and defenses. Similarly, commandos engage in unconventional operations, such as attacks behind enemy lines, to create an advantage at a specific place and time.

Technology may enhance military efficiency and effectiveness. Armor, shields, and stealth systems help preserve military power by reducing vulnerabilities. Long-range weapons enable forces to strike from afar, often preventing an effective counterattack. Advanced mobility systems enable forces to maneuver out of harm's way and to exploit their advantages in positioning and firepower.

Corollary of Maxim 10: Avoid a stronger opponent to preserve military power.

Intangible force multipliers such as leadership, morale, training, and combat experience can enhance a military force's effectiveness beyond raw numbers or firepower. Strong leadership provides vision, discipline, and unity of effort, enabling forces to operate cohesively under pressure. High morale sustains a unit's will to fight, endure hardship, and recover from setbacks. Rigorous training builds competence, cohesion, and confidence, allowing forces to execute complex tasks instinctively and adapt to changing battlefield conditions. Experience, particularly in combat, sharpens judgment, improves decision-making under fire, and fosters the intuitive sense of when to press an advantage or break contact. Together, these qualities can enable a smaller or less technologically advanced force to outmaneuver and defeat a numerically superior enemy.

A savvy strategist can inflict disproportionate damage relative to the effort applied by targeting an enemy's force multipliers or the critical dependencies that support them. As discussed earlier, dependencies such as logistics networks, communications, key technologies, or leadership structures form the backbone of sustained operations. For example, if a fleet consolidates its critical supplies on a single vessel, that ship becomes a point of vulnerability. Rather than engaging a heavily armed fleet head-on, an opponent might strike its weaker logistics ships, disrupt the supply of fuel, food, or ammunition, and achieve strategic effects with minimal effort and risk.

Similarly, concentrating senior political, military, scientific, or religious leaders on a single vessel or at a single meeting creates a high-value target. Capturing or destroying this concentration can decapitate leadership and paralyze the broader organization. Distributing force multipliers across the force mitigates the risk that a single strike will incapacitate the entire structure, as the Rebels did in *Star Wars* when they destroyed the second Death Star and killed the Emperor and Darth Vader.

Corollary of Maxim 10: Concentration may not require collocation. Don't put all your eggs in one basket—your enemy may destroy the basket.

Leaders should preserve force multipliers for the missions to which they are best suited. Force multipliers should be reserved for operations that demand their unique capabilities, not diverted to tasks that conventional forces can accomplish. Assigning them to routine missions wastes their potential and diminishes overall combat effectiveness. Likewise, commanders should seek to neutralize an opponent's force multipliers to prevent those capabilities from enhancing enemy operations.

Corollary of Maxim 10: Conserve strategic advantages.

In *Dune*, Paul Atreides survives the Emperor's attempt to destroy his father and House Atreides on the desert world of Arrakis. He makes contact with a resilient Fremen tribe and is eventually accepted into their ranks. Over time, Paul develops a profound understanding of Fremen customs, values, and way of life, earning their respect and admiration. His extraordinary abilities, combined with their belief that he is the prophesied Kwisatz Haderach, strengthen his influence and transform him into a potent force against the oppressive rule of the Harkonnens and the Emperor. Yet beneath this rise lies a persistent tension. Paul does not seek power for its own sake, and he recoils from the terrible futures his prescient visions reveal. To some, he appears a reluctant savior, acting to save humanity. To others, he becomes the architect of an inevitable holy war. The truth remains a matter of perspective.

Paul soon realizes that uniting the various Fremen clans into a single army is essential to defeating his enemies. To do this, he must first lead his own clan before he can rally the others under one banner. The incumbent clan leader, Stilgar, has accepted Paul into the clan, saved his life several times, and become his trusted friend. However, Fremen tradition requires that any challenge for leadership be resolved through a duel to the death.

Although Paul is the superior warrior, he refuses to kill his friend and mentor. He also recognizes that following the clan's tradition would weaken the clan by eliminating its second-best warrior before a campaign. He persuades the clan to change its custom and accept him as leader without bloodshed, preserving their military power. He declares, "You do not break your blade before a fight" (Herbert 1965).[2]

Corollary of Maxim 10: Don't reduce a force's strength in the name of tradition.

Economy of force emphasizes *proportionality*, using only the minimum necessary force to achieve an objective. Applying excessive force to a low-priority target wastes resources and violates this principle, just as committing insufficient force risks mission failure. Proportionality ensures military efforts remain focused, resource-efficient, and tactically sound while still achieving the desired outcome.

This *restraint* helps minimize loss of life and property damage, preserving infrastructure that may be essential after the conflict. It also conserves limited resources, ensuring their availability for future operations. Excessive force not only squanders those resources but may also raise ethical concerns. Unrestrained violence can alienate populations, erode legitimacy, and undermine long-term objectives, increasing the risk of strategic failure.

Corollary of Maxim 10: Conserve expendables like bullets, bombs, or missiles if durable weapons such as knives or beam weapons accomplish the objective.

Restraint may also build trust with civilian populations and other stakeholders by demonstrating discipline and respect. This can foster cooperation and support, which may be essential for long-term success. Moreover, restraint helps prevent escalation; when one side uses excessive force, the other side is more likely to respond in kind, increasing the risk of uncontrolled violence.

Maxim 26: Greater violence leads to greater resentment. Restraint may reduce resentment and resistance among the conquered, making long-term stability more attainable.

A commander should limit collateral damage from attacks that offer little gain. Such damage often occurs when destructive weapons or tactics are employed, or when military targets are near civilian populations.

The concept of proportionality is frequently invoked to justify or dispute allegations of atrocities. Yet, in the absence of legal or moral restraints, the calculus changes. In existential conflicts, when a society or species is fighting for its survival, the principle of minimizing violence may be abandoned in favor of military expediency at any cost.

CONCLUSION

Massing military power and applying economy of force are two sides of the same coin, both essential to shaping the outcome of military operations. Massing military power underscores the importance of concentrating overwhelming strength, including troops, firepower, or other critical means, at the decisive point of an engagement. This focused application of force aims to overpower the adversary, seize the initiative, and tilt the balance of conflict in one's favor.

In contrast, economy of force recognizes the finite nature of resources and the need to use them judiciously. It involves allocating the minimum essential resources to secondary or less critical operations, enabling commanders to maximize military power at the decisive point. This principle requires prioritization, as not all objectives merit equal importance. It is the art of achieving effectiveness while conserving strength and is a crucial aspect of sound military planning.

Success often hinges on a commander's ability to balance these imperatives: knowing when to concentrate force for decisive effect and knowing when to disperse or withhold resources to preserve flexibility. This balance between overwhelming force and prudent resource conservation represents the hallmark of a skilled strategist.

CHAPTER 10

GEOGRAPHY

"It's over Anakin! I have the high ground."

— OBI-WAN KENOBI, JEDI MASTER (LUCAS 2005)

Just as a politician must understand geography to wield power effectively, a military strategist must understand the environment, often referred to as the *battlespace*, in which they operate. On a planet or other orbital body, this involves understanding how the terrain, weather, flora, and fauna affect combat operations. In outer space, a warfighter must understand the physics of moving bodies and the dangers unique to space.

The battlespace shapes every aspect of military operations. Adverse weather can slow movement, obscure visibility, and conceal troop positions. Open terrain, such as deserts, may facilitate rapid maneuver but often demands more logistical support and exposes forces to surveillance, precision strikes, or area weapons, often turning them into kill zones. Dense environments like jungles and swamps severely restrict visibility, hinder movement, conceal friendly and enemy forces and their activities, absorb the effects of various weapons, complicate resupply efforts, and increase reliance on vertical mobility for movement and support.

Corollary of Maxim 5: Strategy and tactics are always tied to a specific battlespace.

Geography shapes the strategy and tactics of war. Terrain that enables rapid and violent battles often favors the offense, while terrain that restricts movement and limits weapon effectiveness favors the defense. Military leaders must also identify the critical locations within a battlespace to secure an advantage.

KEY TERRAIN

Every battlespace has *key terrain* that provides the group controlling and exploiting it with advantages.[1]

The most important key terrain is Geographic Control Points, which provide observation and engagement advantages, making them difficult to assault and valuable for defense. On planets, high ground such as hilltops or tall structures offers a broader and longer field of view, allows defenders to fire downhill, delays and fatigues attackers, and gives defenders downhill momentum when counterattacking.

However, the value of these points depends on prevailing technology. High ground offers little advantage if an opponent can encircle and isolate occupying forces or employ long-range weapons or weapons of mass destruction, such as nuclear or kinetic planet bombardments, against concentrated positions.

Maxim 27: Control and exploit key terrain.
Terrain is useful if forces can exploit it to their benefit.

In outer space, objects with mass create a gravitational field that pulls other objects towards them. The strength of the gravitational pull depends on the masses of the objects and the distance between them. A *gravity well* is a region of space around a massive object, such as a planet or star, where the gravitational attraction is strong enough to affect the motion of nearby objects.

The key terrain around planetary bodies is the locations where their gravitational forces cancel each other out, resulting in small objects staying in position relative to the other bodies, called the Lagrange points. Since the gravitational forces are in equilibrium, once a spacecraft arrives at a Lagrange point, it can reduce fuel consumption and remain in position, launch vehicles and weapons that can exploit the gravity well, and force opponents and their kinetic weapons to use energy to approach.[2]

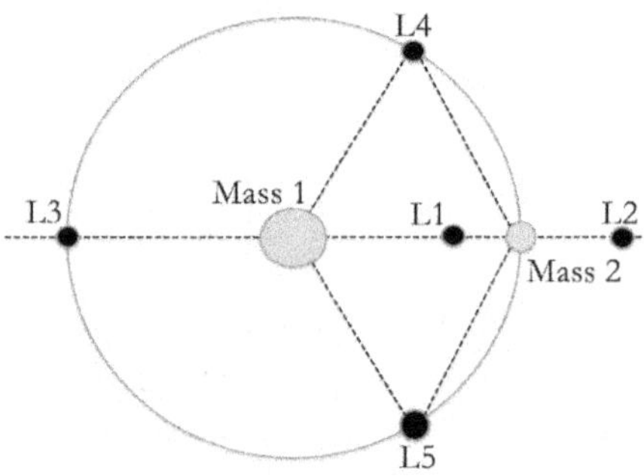

The five Lagrange points of two bodies in orbit around each other.

In *Independence Day,* aliens invade Earth with an orbital mothership and immense City Destroyer spacecrafts that descend to the planet's surface. The aliens destroy waves of human-crewed aircraft that attempt to penetrate the City Destroyers' impervious shields. However, a human scientist intercepts an alien communications signal, inspiring him to take a daring mission. Using an old alien spacecraft, he infiltrates the mothership and infects it with a computer worm. The worm spreads throughout the alien fleet's communication systems and deactivates their protective shields. He then detonates a nuclear weapon aboard the mothership, destroying their alien queen and decapitating the alien invasion force. With the alien force decapitated and their shields down, humanity launches a massive attack against the City Destroyers, defeating the alien spacecrafts and mopping up the remaining alien ground troops (Emmerich 1996).

In this case, the orbital mothership controlled the key terrain and served as the central command-and-control hub for the alien invasion force. Exploiting this key terrain becomes the crux of the human counterstrategy. By infecting it with a computer worm, the humans not only gain a foothold within the enemy's central command but also disrupt the alien communication systems and weaken their overall control.

The City Destroyers, while formidable, are also key terrain, enabling the aliens to exert localized control on Earth. The human-crewed aircraft's initial, futile attempts to penetrate their shields reflect the challenge of confronting well-protected key terrain. However, the City Destroyers' weakness lies in their dependency on the motherships; once humanity eliminates this dependency, the City Destroyers become vulnerable, and humans exploit this power shift.

The scientist's actions against the mothership exemplify the strategy of exploiting vulnerabilities within key terrain. He identifies a weakness in the alien technology and its dependencies, leading to a breakdown in their coordination and defense. It is a critical turning point in the conflict, as it offers humans the opportunity to launch a massive counterattack against the City Destroyers.

Complement of Maxim 27: Losing control of key terrain may yield an advantage to the enemy.

Other key terrain includes transportation points, regions, and paths that enable vessels to travel quickly or secretly. These include choke points that constrain movement, such as valleys on planets, and areas that enable movement, such as highways, hyperspace lanes, and wormholes. Later discussions will show how such locations become prime targets for *interdiction* during military blockades and economic conflicts.

Leverage points are areas where military forces may exploit their opponents, such as on their opponent's flanks and rear. Depending on the environment, these may include the top or bottom of forces, which are vertical flanks. They may also include the seams between forces, exploiting their lack of coordination and their inability or reluctance to collaborate. These points are less about the terrain and more about an enemy's disposition within the battlespace.

Preparing for war involves securing *logistics points*, including ports, refueling stations, and bases. These are political, economic, or military objectives that provide resources, manufacturing, refinement, or maintenance means. Preventing an opponent from accessing logistics points is equally important, achievable through alliances, capture, or destruction of these means.

A point need not provide a physical advantage; a *morale point* refers to a location or idea that provides a psychological benefit to one or both sides. This could encompass the damage, destruction, or capture of political, religious, or cultural centers, as well as control of media outlets, propaganda platforms, key leadership figures, influencers, and strategic communication networks.

Alliance points are areas or concepts that are the basis of a partnership. If an alliance is built on trust, an adversary may try to undermine it. The members of an alliance should try to protect and reinforce the basis of an alliance, such as by using exchange officers (see Chapter Twelve).

INTERIOR LINES

Interior lines are maneuver corridors between forces enclosed within an area. They provide an advantage to the forces operating within the enclosed area against an opponent outside it. An enclosed area often occurs in a *salient* or siege, and in a two-front war. In the diagram below, force A can exploit interior lines to move forces from left to right more quickly and easily than opponent B.

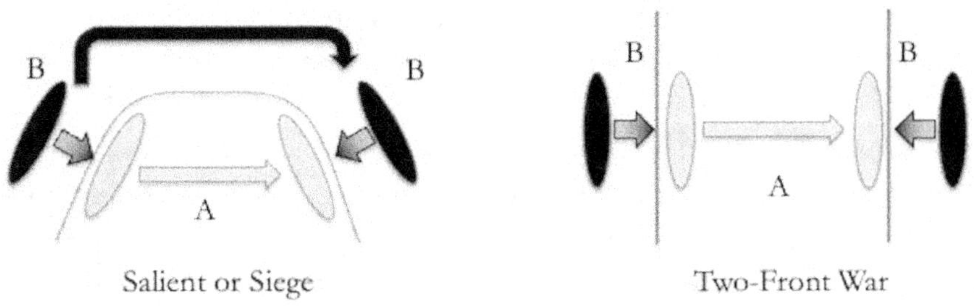

Salient or Siege   Two-Front War

The advantages of the interior lines in the defense (gray).

A force within a salient is typically viewed as being at a serious disadvantage, as an opponent might encircle and annihilate it. However, forces within an enclosed area have shorter lines of movement, enabling them to maneuver with greater safety and speed. This advantage provides the interior force greater flexibility in massing effects and in achieving *local superiority* against a portion of the outer force.

Likewise, a pursuer may cut inside its prey's movement arc to either cut it off using a *blocking position* or maneuver into its flank in an *envelopment*.

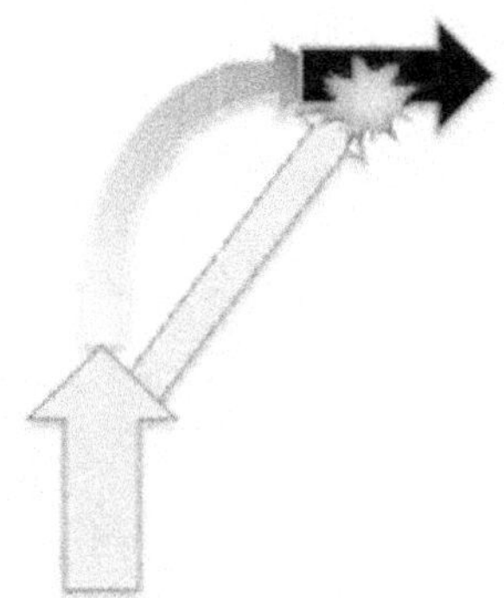

A pursuer (gray) exploits an interior line when their target (black) makes a turn.

Similarly, weapons positioned inside an opponent's movement arc must continue to pass through the weapon's effective arc, allowing the weapon to sustain continuous or repeated engagement. By contrast, a weapon placed outside the opponent's movement arc intersects the enemy's path only briefly. Its engagement is limited to a short window as the opponent crosses the line of fire. Once past that point, the weapon must traverse, reposition, or abandon contact, resulting in a narrower engagement envelope in both time and opportunity.

This explains why defensive weapons are positioned to dominate likely avenues of approach rather than merely cover terrain, and why flanking fires are so effective. They place weapons within the enemy's movement arc rather than simply across it.

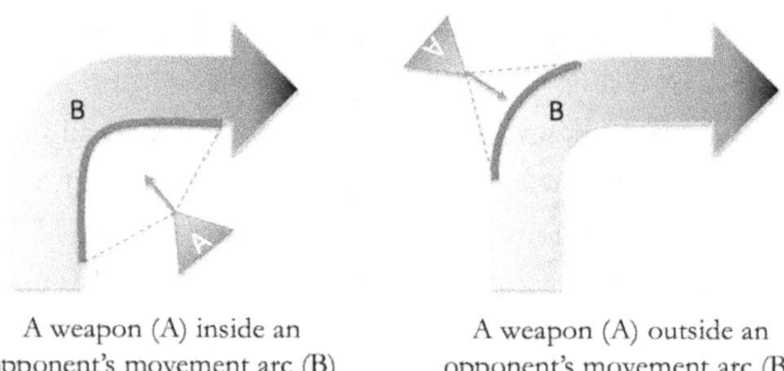

| A weapon (A) inside an opponent's movement arc (B) | A weapon (A) outside an opponent's movement arc (B) |

A tactician can gain temporary advantages by carefully applying the concept of interior lines. When shifting from offense to defense or pausing a maneuver for resupply, a unit may retrograde its flanks to form a salient. This configuration enables concentrated firepower and facilitates resupply by shortening and securing its interior lines.

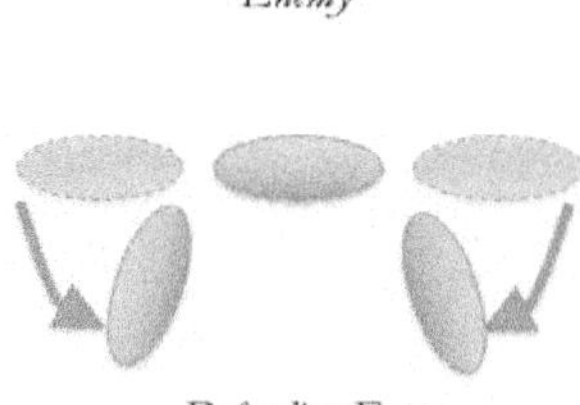

An in-line force can retract its flanks to transition to the defense with interior lines.

GEOSPATIAL INTELLIGENCE

Geospatial intelligence is the understanding and use of geographic information to support military decision-making and operations. It encompasses terrain, infrastructure, weather, population distribution, and cultural nuances. This form of intelligence enables forces to exploit natural features for concealment, predict enemy movement based on terrain constraints, identify efficient supply routes, and navigate cultural dynamics to influence or exploit civilian behavior. Properly applied geospatial intelligence enhances situational awareness and supports political and military planning and operations.

Maxim 11 highlights the critical link between time and space in military strategy: greater distance reduces combat effectiveness. Every additional step drains troop strength, burdens logistics, and gives the defender valuable time to organize, reinforce, and respond.

Corollary of Maxim 11: Battlefield utility is inversely proportional to travel time.
The quicker something can reach its destination, the greater its effectiveness.

This principle becomes even more critical when advanced weaponry and vast interstellar distances introduce unique geospatial challenges. While technologies such as warp drives or wormholes may reduce travel time, constraints such as limited resources and strategic chokepoints will restrict movement. Traversing vast distances and navigating celestial phenomena requires advanced technology, meticulous logistical preparation, and deliberate strategic planning. Planetary environments far away from any assistance may present unforeseen challenges that demand adaptation and improvisation. Moreover, communication delays across interstellar space require proactive planning and decentralized decision-making to sustain effectiveness.

A society may reduce the impact of its movement speed by positioning military forces in forward locations, including border regions, neutral zones, and within allied or conquered territories (Boulding 1962). Technologies capable of delivering weapon effects, such as drones, missiles, and advanced transportation systems, may diminish or even negate the significance of geographical proximity in military strategy.

However, it follows from Maxim 11 that the further that vessels and colonies are from their society's center of power, the less control the central power may apply to them. While a central power may have incredibly fast transportation that enables it to reach the edge of its controlled space, it must depend on sensors, communication and transportation systems, fuel, mechanical and maintenance systems, and mobile forces. The deliberate or accidental damage, capture, or destruction of these dependencies can prevent or delay support to a distant colony or military force. Similarly, changes in transportation technologies may create entirely new strategic powers.

In the **Vorkosigan Saga** series, naturally occurring wormholes enable interstellar travel and link human-inhabited worlds. When these wormholes collapse, entire planets become isolated, severed from trade, communication, and reinforcement until alternative routes are discovered. This disruption has profound effects on political, economic, and military power, triggering intrigue and conflict as factions compete to locate, secure, and control vital wormhole corridors (Bujold 2000).

Similarly, as a society's sphere of influence grows, it may acquire land and resources, but this expansion inevitably thins military power as a fixed-size military force must now be spread out to cover more territory. Depending on the distance and the technology involved, a dispersed and stretched force may lose the ability to observe allies, borders, and enemies, resulting in gaps in sensor networks and limiting the capacity of military units to support one another.

As this society's interior lines extend, so do the supply chains and *lines of communication*. This growth can strain its ability to transport supplies or reinforce units in distress. If the battlespace lacks natural chokepoints that can be held with fewer forces, continued expansion demands a military buildup proportional to the volume of space being added. Without such growth, the society risks exposing critical vulnerabilities across its expanded frontier.

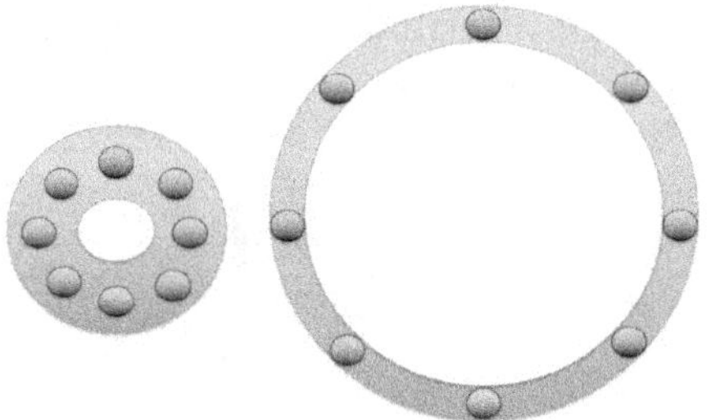

Constant military strength must thin to cover a larger perimeter.

Conversely, when a society contracts, it shortens its interior lines, supply routes, and lines of communication. This can increase local military power and ease the logistical burden of transporting supplies across a region. However, greater concentration may also make a force more vulnerable to weapons of mass destruction and increase the risk of traffic congestion.

Corollary of Maxim 5: Geographic expansion diminishes relative military power, while geographic contraction increases it.

Consider an emperor who has a military force spread equally around a sphere with a radius of 10 parsecs from its capital. Each military unit supports the others in all directions, forming a continuous defensive shell. An invader would face a 10-parsec-deep battle to reach the capital at the center.

Now, the emperor orders their forces to expand outward, doubling the empire's radius to 20 parsecs. This seems like a simple increase, but the results may be catastrophic. Though the distance only doubled, the empire's volume, and thus the territory to defend, has increased by a factor of eight. Unless the military grows proportionally, its coverage shrinks dramatically. Instead of forming a thick defensive belt, the troops may now be spread so thin that they form a fragile shell less than a parsec thick. Any invading force that penetrates this shell finds little resistance beyond it.

Worse, with troops stretched around a vast perimeter, they can no longer support one another. If they rush toward an incursion point at top speed, they will arrive one by one and risk being defeated in detail.

The emperor has turned a formidable defense into a soft-boiled egg: hard on the outside, hollow at the core.[3]

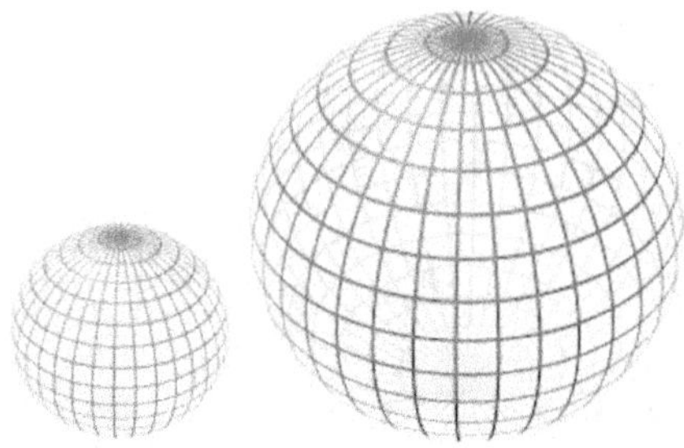

When the radius of a sphere doubles, the volume expands by a factor of eight.

As a society expands, it faces additional challenges, including the OODA loop discussed in Chapter Eight. A distant yet authoritative leader may not observe the battlespace directly, relying on intermediaries to gather, interpret, and report information. The leader must then process this secondhand input through their own OODA loop, which includes observing, orienting, deciding, and acting, before issuing orders to remote forces, who must repeat the cycle before acting. This creates a slower, compounded decision-making process. A more agile opponent may exploit this delay and outmaneuver the cumbersome dual-loop structure. To remain effective, the remote leader may be forced to delegate authority or risk failure.

In **Among Thieves,** humanity has expanded beyond Earth to colonize numerous worlds. Over time, Earth becomes complacent, no longer reliant on its once-powerful and technologically advanced military, instead entrusting its defense to the outer societies. One such society, the Norstad, remains trapped in a perpetual war with the Kolresh, a brutal and cannibalistic warrior race. Both have invested heavily in military arms and training to sustain the perpetual war, while Earth invests in comfort, luxury, and a soft way of life. As the Norstad and Kolresh continue their struggle, they realize they have more in common with each other than with Earth. Together, they recognize Earth as a ripe and unguarded fruit there for the taking (Anderson 1957).

NAVAL FORCES

"We are the Imperial Navy. We do not retreat."

— CAPTAIN GILAD PELLAEON, IMPERIAL NAVY
(ZAHN 1991)

Distance is a relative concept shaped by geography and technology. For some, a desert or ocean may present a formidable obstacle; for others, the vast emptiness of space between planets or star systems becomes an obstacle. The ability to operate within or traverse these natural buffer zones offers a society critical advantages, including:

- Securing access to, and control over, sources of wealth and power along distant routes by protecting commercial shipping, opening new markets, and gaining valuable foreign goods and resources.
- Interdicting an opponent's commerce by capturing or destroying shipping and supply lines, weakening their economic base.
- Establishing a foothold on foreign territory by transporting, protecting, and sustaining invasion forces across great distances.
- Demonstrating a military presence to support political goals, reinforce alliances, or intimidate foreign governments and populations.
- Neutralizing foreign military threats that obstruct any of the above objectives, ensuring freedom of action across extended distances.

**Maxim 28: Project power to increase influence.
Use all forms of power to achieve strategic ends.**

Operating within and beyond buffer zones requires significant investment to establish a navy, whether framed as a reconnaissance or exploration fleet, a colonization or merchant fleet, or an expeditionary force. What distinguishes a navy is its ability to overcome the challenges of these vast zones, achieve its objectives beyond them, and return. Vessels that are lost, damaged, or destroyed may never emerge, consumed by the buffer zone itself. However, if friendly forces can successfully traverse these regions, they gain the ability to threaten an enemy's sources of power while simultaneously expanding and protecting their own. *Force projection* is the ability to deploy and sustain military forces in distant regions, enabling a society to exert influence and respond to threats beyond its borders.

**Maxim 29: Don't strike a foe's home and neglect your own.
Prioritize safeguarding yourself before venturing to threaten others.**

A navy may achieve its unique logistical requirements by making ships self-sufficient, traveling in convoys with support vessels, or establishing logistics points along a route, such as the space stations in ***Babylon 5*** and ***Star Trek: Deep Space Nine***.

A society may secure distant logistics points by seizing them outright or by negotiating alliances and trade agreements to gain access to foreign bases. These outposts may provide critical functions, including crew rest, vessel maintenance and repair, protection, and resupply during or after extended operations. Once established, these bases become *geographic control points*, enabling naval forces to mass defensively or launch offensive operations.

Leaders may discover that crewing vessels deployed for months or years requires personnel with a rare blend of technical skill, endurance, and a deep sense of adventure. The long lead time required to design and build their vessels, the specialized facilities needed for fabrication and maintenance, and the resources and technology required for extended and arduous journeys all constrain a society's ability to establish and sustain a navy. Targeting the sources of these dependencies, such as shipyards, supply chains, or training institutions, may provide an indirect but effective means to cripple an adversary's naval power.

Because of the force projection capabilities of a viable navy, a political leader may view a neighbor's fleet as a threat, not only to the regional balance of power but also to their own political influence, economic interests, military security, or cultural reach. Even if the fleet remains in port, its very existence may provoke anxiety. This perception alone can prompt calls for a preemptive strike, justified by the need to neutralize a rising threat before it matures.

Such decisions, however, carry the risk of significant unintended consequences. A preemptive attack might galvanize the targeted society, triggering nationalist sentiment, alliance activation, or full-scale retaliation. Economic disruption, refugee flows, and shifts in diplomatic alignment may follow. What begins as a calculated move to protect national interests may instead destabilize the broader region, undermine the attacker's legitimacy, and ignite a wider conflict. In this way, the mere presence of a capable navy can reshape strategic landscapes, sometimes by its actions, but just as often by the fear it instills.

CONCLUSION

Military leaders must adjust their means and ways to the unique conditions of each battlespace. Strategies that are effective on the ground, such as halting an opponent and then attacking a flank, may be ineffective in space, where vast distances, open visibility, and the vacuum environment allow adversaries to detect and counter flanking maneuvers well in advance. Similarly, conventional ground forces may be ineffective for operations involving the defense and seizure of spacecraft. Boarding actions demand specialized skills, including precise orbital maneuvering, automated security measures, combat in zero-gravity and vacuum environments, close quarters combat, resource limitations, and communication difficulties. Such missions may be better entrusted to specially trained spaceborne marines.

In the *Halo* universe, the Jiralhanae, known to humans as Brutes, are a fierce, tribalistic alien warrior species that were integrated into the alien Covenant. Though often perceived as dim-witted, they possess spacefaring technology, engage in space-based combat, and maintain an advanced, if brutal, civilization. Capable of cunning tactics and driven by relentless self-determination, they are especially dangerous when provoked, entering a berserk, rage-fueled state when pack members are harmed. Their incredible physical strength, resilience, and ferocity make them very effective ground and boarding troops. However, their aggressive tactics are ill-suited for the long-range precision of space combat, where a disciplined opponent could stand back and pound their vessels into their atomic elements from a distance.

Having defined the battlespace, attention now turns to mastering movement within it. War is inherently chaotic, and effective leaders must transform that chaos into deliberate, forceful motion guided by a defined end.

CHAPTER II

MANEUVER

"I'm a leaf on the wind. Watch how I soar."

— HOBAN "WASH" WASHBURNE, PILOT, SERENITY

(WHEDON 2005)

Maneuver is the movement of forces in relation to an enemy that provides an advantage. Maneuver is more than moving into position to strike an opponent. A commander may maneuver to adjust forces, change direction, seize or create opportunities, shift between offense and defense, create a panic in an opponent, or escape from a threat. Effective maneuver keeps the enemy off balance and protects friendly forces by exploiting successes, preserving the initiative, and reducing vulnerabilities.

Corollary of Maxim 21: Maneuver to provide friendly forces with an advantage and an opponent with a disadvantage.

Maneuvering complicates an enemy's efforts by rendering their actions ineffective, circumventing obstacles, outflanking advances, and evading points of strength. It functions much like a parry and riposte in fencing. Through stance, timing, control of distance, and psychology, a warrior deflects, delays, or neutralizes an opponent's attack, then delivers a decisive counterstroke while the enemy is off balance and unable to recover.

Corollary of Maxim 21: Blunt an opponent's attack and then strike when it is off balance.

People often think of maneuvering as an attack against a defense. However, a defending force can also maneuver. A defense may slow and weaken a larger attacking force, enabling the defender to attack from their defensive positions and defeat the weakened, off-balance, and overextended invader. This *counterattack* can shift the defender from a passive role to an active one, and through it, allow them to resume the initiative.

The way opposing forces encounter each other significantly affects the maneuvers they can employ. In a *chance encounter,* also known as a *meeting engagement,* forces stumble upon each other unexpectedly. This could occur when two units approach each other from opposite directions. Their response to this encounter will often depend on their specific missions. They might attempt to break contact and avoid further engagement, or they could try to hold their opponent in place with minimal forces while maneuvering for an advantage. The key to success in these situations lies in gathering intelligence about the enemy while maintaining flexibility, such as by preserving a reserve force that is free to maneuver.

A *reluctant encounter* occurs when a leader is forced into an undesirable situation, such as being surrounded, flanked, or overrun. A force that stumbles into an opponent or finds itself in an ambush may disengage and withdraw, launch an immediate counterattack, or hold its ground. Regardless of the situation, they must quickly assess it and act decisively to regain the initiative or face potential destruction.

For example, a force caught in an ambush faces destruction. However, applying Maxim 21, "strike when an opponent is vulnerable," and Maxim 25, "when faced with a no-win situation, change the situation," demonstrates that survival depends on turning the tables.

In such situations, the force must assault the ambushers with overwhelming ferocity to seize the initiative and avert annihilation.

Sometimes a force may have time to prepare for an encounter by establishing standoff weapons, booby traps, and defensive positions that delay, weaken, disrupt, and demoralize its opponent. If withdrawal becomes necessary, the force may deploy a rear-guard force to hold the line and buy time for the main body to withdraw in good order.

In *The Lost Fleet: Valiant*, the Alliance employs deception as a force multiplier by planting booby traps aboard abandoned and damaged Alliance vessels and captured enemy ships. These hulks are then deliberately positioned to appear as derelicts left behind in the aftermath of an earlier battle, inviting investigation or recovery by the advancing Syndicate fleet. Through careful maneuvering, the Alliance draws the pursuing ships into proximity with these seemingly harmless wrecks. Once in position, the vessels' power cores detonate, turning the vessels into massive spaceborne mines. The blasts destroy many vessels and inflict severe damage on nearby ships, shattering formation cohesion, and creating confusion in the enemy command structure. Beyond the immediate physical destruction, this tactic may undermine the pursuers' confidence, forcing them to question whether every future drifting ship conceals a hidden threat, slowing their advance, and fracturing their morale. Such psychological and material effects exemplify how a well-prepared battlespace can delay, weaken, and demoralize an opponent before decisive engagement.

This optimal situation is the *set-piece battle,* sometimes called a *pitched battle*, in which a leader "sets" forces on terrain to maximize their strengths against an opponent's weaknesses. The leader makes the most of the battlespace geography, maneuver, intelligence, and deception to defeat their opponent. The most common techniques for set-piece battles are penetration, infiltration, envelopment, turn, and swarm.

PENETRATION

A penetration, also known as a breakthrough, involves a powerful force piercing an enemy line and forming a breach. Leaders typically use two forces: a holding force that maintains the breach and keeps the defenders in place, and a penetrating force that drives through the breach and continues deep into the enemy's rear area.

The penetrating forces may exploit their success by turning back and attacking the enemy from the rear, or by proceeding to destroy the enemy's command and control, secondary forces, and logistics elements.

The holding force may conduct secondary attacks along an enemy line to fix the opponent and prevent its maneuver. However, commanders must balance the need to retain sufficient power to penetrate the enemy line, overwhelm defenders, and exploit any breach against the risk of diverting too many forces into secondary attacks that merely hold the enemy in position.

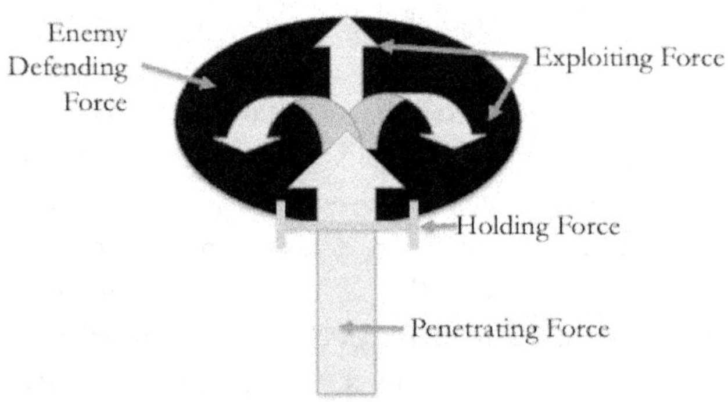

A gray force penetrating the black force, as seen from above.

Success in a penetration often requires an invader to:

- Suppress enemy forces to reduce their ability to target the penetrating force.
- Obscure the battlefield so enemy gunners and weapon

systems cannot observe or effectively target penetrating forces (e.g., by using smoke, electronic jamming, or chaff).
- Penetrate and secure a path at the breach point through the enemy forces.
- Exploit the penetration.

A frontal attack into the center of an opponent's forces is a high-risk maneuver, since the defender may concentrate all its weapons against the attacker. To reduce this risk, attackers may search for seams between units that represent weaker points of penetration. However, even a successful breach will leave the penetrating force surrounded by the defenders. It's crucial that the attacker maintain the initiative by continuing to maneuver and by keeping pressure on the defender, or risk being enveloped and destroyed.

Inverse of Maxim 21: A frontal attack targets an opponent's strength.

INFILTRATION

"We're not going to the border; we're going through it."

— MALCOLM REYNOLDS, CAPTAIN, SERENITY
(WHEDON 2005)

Where a penetration is a visible, often violent, piercing of enemy defenses, an *infiltration* is the surreptitious movement of a small force into or through enemy-held territory with the intent of avoiding detection and engagement. An infiltration may function as a maneuver within a larger operation or an operation in its own right. In either case, it enables access, positioning, or effects that would be difficult or impossible to achieve through overt means, and it commonly supports penetrations, raids, ambushes, and harassment activities.

An infiltration is a common tactic used by smaller forces against larger ones, especially in guerrilla operations. Agents inserted into an enemy's territory or organization can gather intelligence, disrupt communications and supply lines, and conduct sabotage. This can help them level the playing field and gain an advantage against a more powerful adversary.

The Rebel Alliance in *Star Wars* frequently employs infiltration missions to undermine the Galactic Empire's efforts. In *Rogue One: A Star Wars Story,* a group of rebels penetrates the planet Eadu to save or assassinate the Imperial scientists who are building the Death Star battle station (Edwards 2016). Later, they infiltrate Scarif to recover intelligence on a secret vulnerability in the Death Star. In *Star Wars - Episode VI: Return of the Jedi,* rebels infiltrate the Imperial base on the moon of Endor to disable shield generators so they can destroy the second Death Star (Lucas 1983).

An infiltration may be used to:

- Sneak something or someone, such as intelligence agents, saboteurs, or non-combatants, into, out of, or through a combat zone.
- Gather intelligence about an enemy or terrain.
- Link with other friendly forces or allies.
- Seize key control points before a battle.
- Conduct raids, ambushes, assassinations, or designate targets for long-range strikes.
- Breach a path to allow larger forces to enter or exit enemy territory undetected.

Infiltrators often must operate independently when they penetrate enemy lines, discover routes, bypass opponents, select and engage targets, and function within enemy territory. However, other units may assist the infiltration by creating diversions to hold, suppress, or distract opponents.

Operating independently in enemy territory poses significant risks, requiring infiltrators to focus heavily on their own security. They may use disguises, such as wearing enemy uniforms or using captured vessels with valid identifying signals. Since their disguise can be effective, infiltrators may use procedures and recognition signals to prevent friendly forces from mistaking them for the enemy during infiltration, exfiltration, or operations behind enemy lines.[1]

While infiltrations are a favorite technique with irregular forces, powerful forces may also employ them. In the *Terminator* universe, the Skynet self-aware artificial intelligence system employs increasingly complex and realistic Terminators disguised as humans. These assassins infiltrate human bases and teams, and once inside, massacre combatants and non-combatants alike.

A strategist can employ infiltrations for more than their physical effect. They may serve as reconnaissance by fire, provoking a response that reveals an enemy's capabilities or disposition. Surprise attacks can generate terror and confusion, creating profound psychological effects on victims, onlookers, and supporters. Such operations can reshape a society's political landscape by portraying the aggressor as liberator, protector, or avenger, while casting doubt on the defender's competence. Repeated, unexpected raids can instill fear and uncertainty, leading citizens to question their safety and their government's ability to protect them. This erosion of trust can incite civil unrest or demands for greater security measures. Each successful incursion sends a powerful message: "We breached your defenses; your leaders failed to protect you." The result is a weakening of political credibility, the loss of allied confidence, and the emboldening of adversaries.

Successful infiltrations may foster suspicion within the targeted society, turning citizens into suspects. In response, the defender may increase surveillance, harden defenses against future attempts, or launch reprisal strikes against suspected collaborators. These responses may inadvertently play into the attacker's strategic plans.

ENVELOPMENTS

"It's a trap!"

— ADMIRAL ACKBAR, ADMIRAL, REBEL ALLIANCE
(LUCAS 1983)

A flanking attack, also known as a single envelopment, is a convergent assault that often uses a holding force to keep an enemy in position while an assault force attacks the enemy's weaker flank. After successfully penetrating or destroying the flank, the assault force may attack toward the rear to threaten the opponent's logistics, communications, lines of retreat, and command and control, or destroy the defenders in place. If the opponent turns to address the flanking attack, the holding force may launch a converging assault against the enemy.

In most flanking attacks, the attacker must maneuver quickly to the enemy's flank before or as it is exposed, then attack into its core or rear area. For this reason, attackers often use highly mobile forces for their assault force.

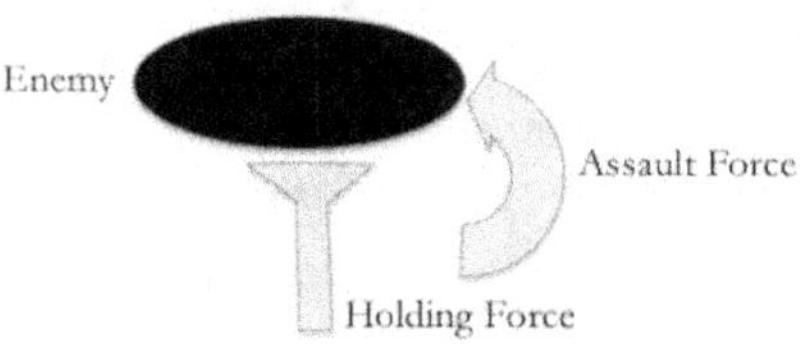

A flanking attack (gray) as seen from above.

A pincer, also known as a double envelopment, uses two simultaneous assault forces to crush an opponent in a vise.

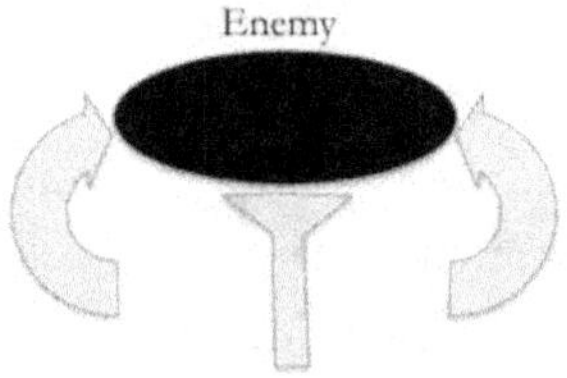

A pincer attack as seen from above.

A defender may lure an enemy into a special pincer operation, known as a cauldron battle.[2] By making their center appear weak or intentionally collapsing it, they form a "V" shaped salient that draws their opponent deeper into their territory. The defender's center force becomes the holding force, and assault forces on the two flanks close and crush the attacker's flanks.

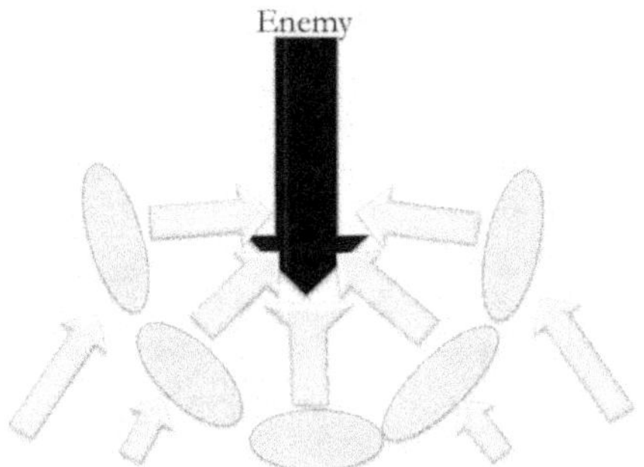

A cauldron battle catching an attacker, as seen from above.

The purpose of an envelopment is to trap and destroy an enemy. When forces discover they are surrounded, they may panic and attempt to flee. However, an encircled foe might choose to fight to the end. A shrewd commander may deliberately leave an avenue of escape or surrender, depending on their broader objectives, to encourage retreat or capitulation rather than provoke desperate resistance.

Enveloping an invader and preventing its escape is most effective when the invader has advanced beyond a safe limit (see "culmination point" in Chapter Seventeen). To create this situation, the defender may contract away from the attacker, then cut off the attacker's escape until the attacker surrenders or dies.

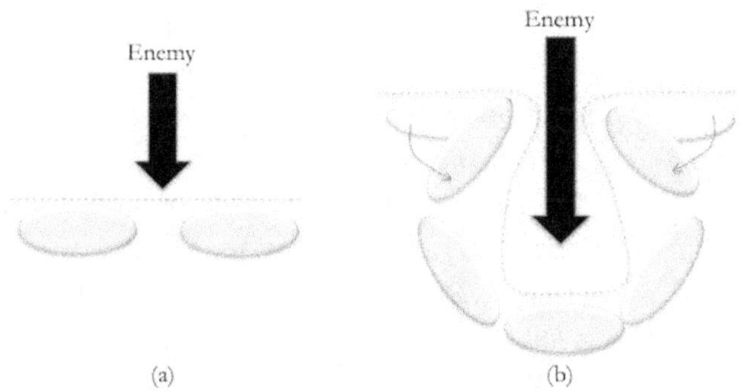

An attack drives between the seams of two defenders (a) and is enveloped in a salient (b).

A vertical envelopment is a flanking attack from above or below, often used to seize key terrain, block an enemy's escape or reinforcement, or disrupt enemy movements.

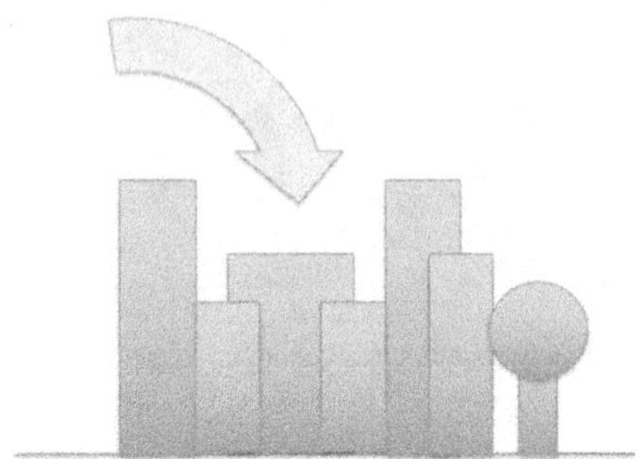

A vertical envelopment of a city, as seen from the side.

The **Starship Troopers** novel opens with a vertical envelopment raid: human starships deploy armored shock troopers directly into the atmosphere of an enemy planet, where they descend on one of its cities. Equipped with jump jets, the troopers maneuver through the urban terrain in *skirmish* lines, unleashing death and destruction in their wake. Their rapid, unpredictable maneuvers prevent the planet's inhabitants, the Skinnies, from fixing their position or concentrating forces against them. The sheer devastation of these hit-and-run urban assaults shocks the Skinnies and compels them to the negotiating table (Heinlein 1960).

Attacks from above can be devastating, but greater surprise may come from directly beneath an opponent's feet. The alien Formics invaded Earth a century prior to the events of **Ender's Game**. They use large colonizing vessels to conduct vertical envelopments on Earth, and after landing, deploy domed shields that prevent humans from striking them. Humans devise a tactic using self-propelled drill sledges, or Tactical Earth Burrowers, to tunnel underground and bypass enemy defenses through vertical envelopment from below (Card and Johnston 2014).

Both examples illustrate how assault forces used in envelopments often find themselves isolated behind enemy lines and vulnerable to being overrun. To preserve shock troops, a strategist must capitalize on the element of surprise and act swiftly to prevent their destruction.

In future wars, it will be especially important that leaders think in three dimensions. This may be a challenge for those who have only fought in two dimensions. In **Star Trek II: The Wrath of Khan**, Rear Admiral James T. Kirk, on the starship *Enterprise*, fights the genetically engineered Khan Noonien Singh on the hijacked starship *Reliant* in the Battle of the Mutara Nebula. After Khan badly damages Kirk's ship, Kirk taunts Khan to enter an interstellar dust cloud known as the Mutara Nebula, aware that the nebula's ionized gas will deactivate both vessels' protective shields and badly degrade visual and tactical systems.

Having captured a powerful new technology and a functional starship, Khan holds all of the military advantages and has achieved his objectives. With Kirk and the Enterprise disabled, he could easily withdraw and achieve his ends at his leisure. However, provoked by Kirk and driven by a personal vendetta, Khan pursues Kirk into the nebula, jeopardizing his apparent victory.

With Khan's advantages eliminated, the two ships play a game of cat and mouse in the nebula, further damaging both vessels. Kirk's science officer, Commander Spock, observes that Khan is intelligent but inexperienced, displaying "two-dimensional thinking" by only attacking on the same plane. Khan's experience on planets has taught him to attack his opponents head-on or from a flank. With the means now even, Kirk's experience operating in three-dimensional space and playing three-dimensional chess provides an advantage over Khan's two-dimensional thinking.

Kirk listens to his subordinate's advice and orders the *Enterprise* to "come to a full stop" and then descend vertically relative to the *Reliant*. When the *Reliant* passes "over" the *Enterprise*, Kirk orders an "ascent" in a vertical envelopment that catches the *Reliant* from behind to deliver a brief but devastating close attack that defeats Khan (Meyer 1982).

TURNS

A turning maneuver involves swinging troops wide around an enemy's position rather than attacking a flank. It often threatens an opponent's supplies, lines of communication, or bypasses the enemy altogether. If the foe turns to engage the threat, it may expose its former front, now a flank, to attack.

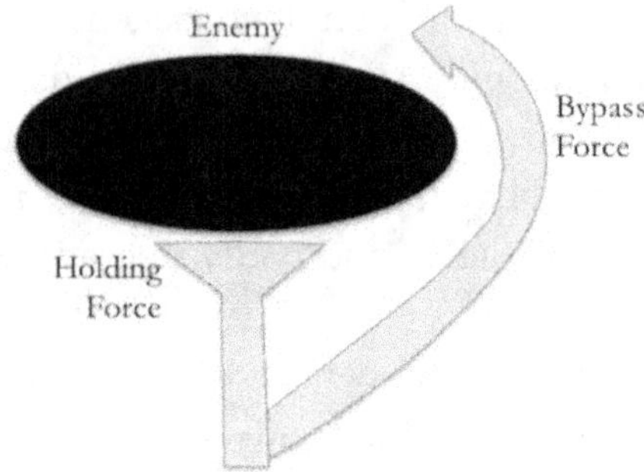

A turning maneuver as seen from above.

An encirclement involves surrounding, isolating, and disrupting an enemy's supplies and reinforcements. Friendly forces can then attack or starve the beleaguered enemy until it surrenders or dies. For more on blockades and sieges, see Chapter Seventeen.

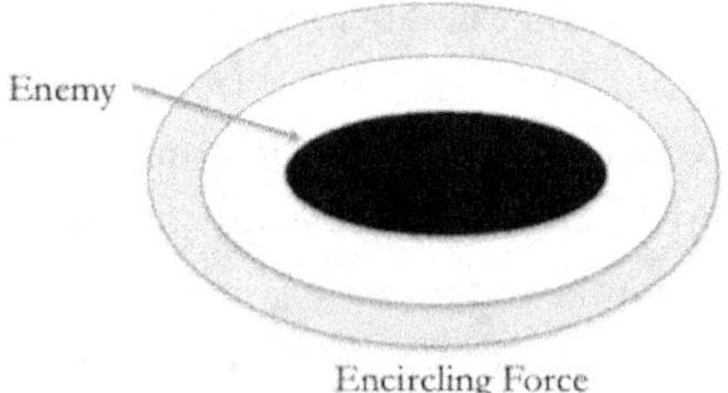

An encirclement, as seen from above.

SWARM

An omni-directional attack, or swarm, involves many small units striking simultaneously from multiple directions. Individually, these units are weak, simple, and inexpensive, but they remain dispersed, presenting few viable targets. Their effectiveness lies in rapid convergence, massing firepower to overwhelm opponents in a surprise maelstrom, then dispersing before an effective counterattack can be mounted.

Like a pack of small, ferocious beasts, a swarm creates chaos through simultaneous and unpredictable attacks. By saturating a larger force's ability to detect, track, and engage threats while defending itself, the swarm overwhelms a more powerful opponent before vanishing.

Corollary of Maxim 21: Swarms create chaos. Exploit this chaos.

On the offense, a swarm may serve as an advance party or shock troops to breach enemy defenses and keep the breach open, or to seize key terrain. On the defense, a distributed swarm might conduct surveillance over a large area to identify hostile targets, molest isolated individuals, and mass to destroy an attacker's main body.

The Rebel fighters (e.g., X-wings) of the *Star Wars* universe often act as swarms when they attack large capital ships and battle stations. The Imperials designed their vessels to defeat other capital ships, and they struggle to defend themselves from small and fast fighters.

A force may eventually defeat a swarm by attacking individual drones. However, a more effective tactic might be to bait and then strike a swarm with area-denial weapons, such as an enormous explosion, an electrical discharge, or an electromagnetic pulse (EMP).

In *Ender's Game,* the International Fleet develops a Molecular Disruption Device to destroy a Formic swarm. This weapon breaks atomic bonds on contact, triggering a chain reaction that spreads outward and obliterates anything in its path. It turns the swarm's greatest strength, its dense formation, into its greatest weakness. As each Formic disintegrates, it becomes a new catalyst, causing the destruction to cascade through the tightly packed mass, turning a previously impregnable wall of bodies into the agent of its own demise.

An alternative strategy may be to attack a swarm's dependencies. The individuals within a swarm often require sensors and secure communications to coordinate their assembly and dispersion, rapid maneuvers, and attacks. A strategist might target the vessel or the control center that directs a swarm, or jam its navigation or communications.

While stranded in the uncharted Delta Quadrant, the crew of the **Voyager** encounters a previously unknown alien species. Unable to communicate and inadvertently trespassing into their territory, Voyager comes under a coordinated swarm attack. Though each alien craft fires only a small amount of energy, their combined assault overwhelms the ship's defensive shields.

With the *Voyager*'s shields down, the swarm members latch onto the vessel's hull and begin draining the ship's energy. The *Voyager* defeats the swarm by attacking the energy lattice it uses to communicate, destroying some drones and forcing the rest to depart (Sussman 1996).

Similarly, in **Star Trek: Beyond**, Krall deploys a fleet of manned swarm ships to overwhelm larger, heavily armed Federation starships. Though individually small and lightly armed, the swarm vessels are heavily armored and operate under a unified directive to inflict coordinated destruction. Their strength lies in synchronized movement and massed attacks. The Federation crew ultimately exploits this dependence on coordination by using high-frequency radio interference, delivered through a burst of "classical" music, to disrupt the swarm's control and cohesion, transforming its greatest strength into a decisive vulnerability (Lin, Pegg, and Jung 2016).

CONCLUSION

Maneuver is the art of moving forces in relation to an enemy to gain an advantage. It is used to escape from enemy strengths and to create and exploit opportunities. Maneuver is often one of the most powerful weapons in war, as it can enable a weaker force to defeat a stronger one through speed, agility, and deception. When used properly, it can introduce physical and psychological effects on the enemy, such as creating dilemmas, disrupting cohesion, and inducing shock. The next chapter will delve deeper into the psychological aspects of war.

CHAPTER 12

PSYCHOLOGY

"Deep in the human unconscious is a pervasive need for a logical universe that makes sense. But the real universe is always one step beyond logic."

— PRINCESS IRULAN CORRINO, WIFE OF EMPEROR
PAUL ATREIDES (HERBERT 1965)

A symbiotic relationship exists between a leader and their subordinates. The leader bears responsibility for fostering cohesion, discipline, and purpose within the unit. In turn, the military force serves as the instrument through which the leader's vision is realized.

Maxim 30: Troopers are a leader's proxy.
Leaders inspire and empower their troops to achieve their objectives.

War, unlike any other profession, tests the psychological core of a leader. A commander burdened by confusion or stress may produce convoluted plans, leaving subordinates disoriented and ineffective. In contrast, clear and decisive leadership yields coherent strategies and focused execution, instilling purpose and direction throughout the ranks.

However, war is chaos. Military leaders must choreograph a complex dance, maneuvering forces, sustaining logistics, striking opponents, and withdrawing and reconstituting troops and materiel, all while facing a thinking adversary intent on disruption and destruction. This struggle unfolds in an environment that is never static. Conditions, assumptions, and relationships shift continuously as actions generate reactions, friction accumulates, and new variables emerge without warning. No plan remains intact for long, and uncertainty compounds faster than it can be resolved.

Unforeseen problems, often referred to as the *friction of war,* obstruct even the best-laid plans. Movements are delayed, battles evolve unexpectedly, and equipment failure, weather, bad luck, and coincidence shape outcomes in decisive ways. Leaders must also contend with the uncertain performance of their forces under stress and with unpredictable shifts in the political environment. In such conditions, misjudgments carry immediate consequences, jeopardizing not only the mission but the lives and cohesion of the force itself.[1]

While no leader can eliminate friction entirely, its effects can be mitigated. Simplifying plans and tasks, maintaining mental and operational flexibility, and sustaining troop morale are essential to preserving combat effectiveness amidst the disorder of war.

SIMPLICITY

From the outside, military operations may seem complex and overwhelming. Yet for those involved, clarity is essential. Leaders must ensure their subordinates fully understand their instructions. Simplicity enables exhausted, stressed troops under fire to make effective decisions faster than the enemy can react.

Maxim 31: Simplicity enables effective decisions.
Clarity and speed in complex situations come from keeping things simple.

As discussed in Chapter Eight, clear command structures in large organizations promote efficiency by establishing well-defined lines of authority. Likewise, simple and unambiguous orders help subordinates grasp not only directives but also a commander's broader intent. This clarity enables them to respond effectively and remain aligned with the mission, even in unforeseen situations.

Leaders can further reduce chaos through doctrine and processes executed by proficient, well-trained troops equipped with interoperable systems. Interoperable equipment ensures seamless support and resource sharing during unforeseen failures. A common doctrine and battle drills create a shared language and knowledge base, enabling swift, coordinated action across diverse units. Standardized processes, ingrained through rigorous training, produce instinctive team responses under pressure. Together, these measures synchronize effort, maximize effectiveness, and strengthen a force's collective strength and resilience in the face of challenges.

Collaboration among diverse cultures, societies, and species presents unique leadership challenges. Each group carries its own communication style, decision-making process, and underlying values. Leadership that appears decisive and direct in one culture may be seen as abrasive or disrespectful in another. Effective leadership in such environments requires adaptability, cultural awareness, and the ability to bridge differing expectations.

Two groups may swap *liaisons* or *exchange officers* to reduce, though not eliminate, complexity and to build trust. These individuals function as cultural bridges within allied forces, fostering mutual understanding. By grasping the nuances of both cultures, they help ensure accurate transmission of information, orders, and intent. Through ongoing interaction and shared experiences, these officers may cultivate trust and respect between allies. They also provide valuable cultural context, enabling leaders to better understand one another, adapt their approach, and collaborate more effectively across diverse groups.

In the *Star Trek* universe, the Vulcans assign Commander T'Pol to Starfleet as an exchange officer. She serves on the *Enterprise (NX-01)* as the science officer and first officer, but retains her rank and title within the Vulcan High Command. Likewise, Starfleet assigns the human Commander William Riker as an exchange officer aboard the Klingon Bird-of-Prey IKS *Pagh* (Haight, Amos, and Armus 1989).

Conversely, Commander Spock, born on Vulcan to a human mother and Vulcan father, fully embraces Vulcan culture and logic despite his mixed heritage. He becomes a prominent member of Starfleet while retaining his Vulcan citizenship and identity. In contrast, Worf, though genetically Klingon, was raised by humans and serves as a Starfleet officer. These examples illustrate how a society can embrace individuals from diverse backgrounds to expand its pool of talent and experiences. However, a society may need to address conflicts arising from its members' different values and cultural backgrounds.

Just as a leader can act to reduce chaos in their forces, they may weaponize chaos and complexity in an opponent's mind to overwhelm and confuse them. This can involve striking at unexpected locations and times, then hitting again before an enemy can recover, disrupting their ability to maintain an OPTEMPO. Likewise, a leader may generate political chaos by decapitating a society's leadership structure (see Chapter Twenty). Removing key leaders may create a power vacuum, destabilize decision-making, and spread confusion throughout the enemy's ranks.

Converse of Maxim 31: Increase the complexity in an opponent's mind to overwhelm and confuse them.

FLEXIBILITY

"Old ideas don't win wars."

— MAZER RACKHAM, ADMIRAL, INTERNATIONAL
FLEET (CARD AND JOHNSTON 2016)

A plan is a theory for shaping and executing a situation to achieve a desired objective or end. Regardless of how well-conceived it may be, its success depends on imperfect individuals, equipment, and devious enemies behaving as expected. Leaders often discover that inaccurate information, unanticipated variables, and the chaotic nature of the battlefield can render meticulously crafted plans ineffective. As a result, blindly relying on these plans can lead to failure. Leaders may realize that "plans are useless, but planning is indispensable."[2]

Maxim 32: Flexibility is strength; predictability is weakness. No plan, no matter how brilliant, survives contact with the enemy.[3]

This does not imply that leaders should refrain from formulating plans. Rather, they should wargame their plans before an operation by contemplating and rehearsing intended actions, anticipating enemy responses, and preparing countermeasures, while deliberately guarding against *confirmation bias* and *mirror imaging*. Effective planning requires challenging one's own assumptions and resisting the temptation to assume an adversary will think, value, or react in familiar ways. In **Battlestar Galactica,** Starbuck warns Apollo that his plan to destroy an installation is "textbook-perfect," and therefore doomed to fail precisely because it assumes a predictable enemy response. Her critique highlights a central lesson of wargaming: plans that appear sound in theory often collapse when they rest on unexamined assumptions about adversary behavior. By stress-testing plans against adaptive and unconventional responses, leaders improve their ability to adjust execution as reality diverges from expectation, preserving flexibility in a dynamic battlefield environment (G. A. Larson 2009).

Beyond planning, a leader must maintain the ability to maneuver forces in response to changing conditions. Reserve forces play a crucial role by allowing a commander to counter enemy actions or exploit emerging opportunities. By prioritizing both maneuverability and the retention of uncommitted reserves, a leader maintains the flexibility to respond to the unexpected and seize decisive moments, increasing the likelihood of securing victory in the face of unforeseen challenges.

Leaders may develop a library of moves and countermoves, relying on well-rehearsed drills to pursue a long-term strategy. This approach can be effective against brief conflicts or against static opponents. Yet training and repetition, while building proficiency, can also breed predictability. Even the strongest force must adapt, or risk being studied and outmaneuvered by an opponent who tailors tactics and technologies to exploit its weaknesses.

Corollary of Maxim 32: Rigid military plans and technologies make military forces predictable and exploitable.

In *Ender's Game,* Andrew Wiggin gains the advantage by embracing adaptability, while his opponents falter through their inability to evolve. At Battle School, students train against static artificial intelligence (AI) opponents and become predictable by relying on the same tactics that succeeded against the unchanging AI. Wiggin also trains against the AI but discovers that "well-rehearsed formations [are] a mistake. It allows soldiers to obey shouted orders instantly, but it also [means] they [are] predictable" (Card 1999). Such rigidity limits effective response to the unexpected. Rejecting this model, Ender cultivates flexibility, relying not on memorized responses to observed stimuli but on creativity and intuition to devise imaginative strategies that defeat both artificial and human adversaries.

He trains his subordinate leaders to exercise initiative and to operate independently to achieve their missions. He forms units that experiment with new and innovative ideas, tactics, and formations, encouraging innovation over routine and drill. This adaptability allows him to anticipate, outmaneuver, and ultimately defeat the alien Formics.

In Jack Campbell's *The Lost Fleet* series, John Geary leads a battered and outnumbered interstellar fleet on a prolonged retrograde out of enemy territory. Geary's success lies in his ability to outmaneuver and outwit his relentless pursuers. Eventually, he encounters an opponent who attempts to imitate his earlier tactics, assuming that replicating his methods will yield similar victories.

However, Geary's prowess as a military strategist lies in his ability to adapt. He changes his approach with each engagement, making his next move difficult to predict. Anticipating that his past tactics might be mimicked, Geary develops and rehearses countermeasures specifically designed to exploit imitators. His foresight and flexibility make him a formidable opponent against those who rely on imitation rather than innovation.

Corollary of Maxim 32: Establish counterplans to deal with a mimic.

Flexibility is especially important when a leader lacks reliable intelligence about an opponent. In such situations, decisions are made under uncertainty, where assumptions may be incomplete, misleading, or entirely absent. Rigid plans, no matter how well rehearsed, collapse when the enemy acts unpredictably or when the environment shifts in unforeseen ways.

To remain effective under these conditions, adaptable strategies become essential. Maintaining a reserve preserves freedom of action, which is critical when the enemy's strength or intent is unclear.

Other adaptive practices include using deception, feints, and misdirection to confuse or fragment the opponent's decision cycle. These actions not only mask one's own intentions but also force the adversary to respond rather than act. By keeping the initiative and remaining unpredictable, a leader can manipulate tempo and shape the engagement, compelling the enemy into a reactive posture.

Ultimately, flexibility allows a leader to test, probe, and learn. It keeps the enemy guessing, buys time for observation, and ensures that the lack of intelligence does not become a fatal weakness.

A military force's adaptability declines as its leadership grows more authoritarian. Autocratic leaders centralize decision-making, issue directives unilaterally, and demand unquestioning obedience. While this approach may yield swift action, it fosters rigidity, suppresses initiative, and stifles creative problem-solving, ultimately impairing their ability to adapt to evolving battlefield conditions.

Alternatively, a leader may decide to give trained and trusted subordinates latitude in executing a plan. They may delegate authority and encourage adaptability when subordinates find themselves in situations that do not align with the plan. However, subordinates must ensure their actions support their commander's vision and that they understand the limits imposed on them. For example, during an attack, does the commander want to focus on seizing key terrain or on destroying enemy forces occupying it? The method subordinates use to achieve their mission will depend on their understanding of this constraint, which is called the *commander's intent.*

Corollary of Maxim 32: Subordinates must understand their commander's intent.
Flexible warriors may adapt to the ever-changing nature of war.

Military leaders who prioritize flexibility and adapt their plans to a dynamic, unpredictable battlefield are better equipped to respond to unexpected developments and adjust tactics on the fly, even when facing situations not explicitly covered by orders or plans.

Understanding a commander's intent is especially important for forces operating remotely, such as expeditionary forces, navies, and guerrillas. They must understand and independently pursue their superior's strategic ends and objectives, since time, security, and communication technology may not allow these units to request clarifications from superiors once deployed.

This isn't to imply that leaders should grant subordinates unlimited freedom to pursue their missions. A leader should not give untrained or untrusted forces leeway; instead, they should require them to follow instructions. Likewise, if a plan absolutely relies on some units succeeding in a certain way, a leader may demand absolute obedience to an order. For example, if an entire battle hinges on a unit seizing a critical objective by a certain time, then the leader must ensure their subordinates understand that accomplishing that mission is essential and that no flexibility is allowed.

A military force's resilience, its ability to withstand and recover from adversity, is essential to long-term success. It includes strategies that enhance the force's capacity to endure challenges and adapt under pressure. Some methods focus on avoidance, such as camouflage or skillful maneuvering to evade harm. Others involve active defense, including shields or reactive armor systems that deflect or absorb threats like space debris, asteroids, or enemy projectiles. Vessels may also be engineered to sustain damage and remain operational, incorporating features such as ablative armor that dissipates energy by chipping or vaporizing under impact.

Engineers may reinforce ships with strong internal frameworks or isolate damage using airlocks and blast-proof bulkheads. Redundant systems, though costly, ensure continued operation even if a component fails. Recovery may involve repair and regeneration, with damage control teams, automated systems, or medical bays restoring function to personnel and equipment.

Over time, resilience may take the form of adaptation to the environment, the enemy, or the mission, culminating, in extreme cases, in the evolution of physical or cognitive traits to address emerging threats.

If failure is the best teacher, then the manufactured Cylons of *Battlestar Galactica* possess a highly effective education system. When a Cylon is "killed," its consciousness, knowledge, and memories are instantaneously transmitted to a new body at a resurrection facility (G. A. Larson 2009). Each death becomes a learning opportunity, enabling analysis of failures and refinement of tactics. Capital punishment offers no deterrent, as death merely returns the Cylon to safety. Even suicide, rather than being an act of despair, serves as a method of escape and of sharing gathered intelligence.

Defeating such a resilient enemy demands creativity. Indirect approaches to defeating the Cylon culture might involve:

- Resurrection Hubs: Destroying or disabling resurrection facilities to eliminate the ability for fallen Cylons to resurrect. This would require accurate intelligence and precision long-range strikes.
- Consciousness Transmission: Interrupting the transmission of consciousness, through jamming or creating

communication dead zones, can trap dying Cylons permanently in death.

- Psychological Disruption: Exploiting psychological vulnerabilities by inflicting trauma before death may degrade their cognitive function upon resurrection, disrupting learning and demoralizing the collective.

Each approach carries its own ethical considerations and potential consequences. Carefully weighing the risks and benefits is crucial when confronting such unconventional adversaries.

MORALE

"Come on you apes! You want to live forever?"

— JEAN RASCZAK, MOBILE INFANTRY (HEINLEIN 1960)

Warfare between living beings is a personal endeavor, with warriors often grappling with the fear of both immediate and future confrontations. To counter this, a leader must inspire troops to fight. Individual morale, however, remains challenging to predict in war. The following factors can influence the morale of troops:

- Volunteers and Morale: Volunteers often enter with a stronger sense of purpose and commitment than conscripts, boosting initial morale.
- Camaraderie: Living, training, and fighting together as *cohorts* builds bonds that strengthen trust, cohesion, and morale.
- Belief in Mission: Soldiers who believe in their cause display greater resilience than those fighting without conviction.
- Defense of Homeland: Protecting one's homeland or loved ones inspires a deeper commitment than fighting abroad.
- Moral Motivation: Fighting perceived evil or injustice can

drive greater determination than personal gain or abstract ideals.

- Training and Support: Well-trained, well-equipped, and well-supplied troops sustain confidence and morale longer.
- Impact of Casualties: Heavy losses, especially in close-knit units, rapidly weaken morale through grief and vulnerability.
- Fear from Weapons: Terror-inducing weapons such as tanks, snipers, incendiaries, or WMDs can erode morale by instilling fear.
- Leadership and Care: Soldiers remain motivated when cared for by officers, valued by leaders, and supported by society.
- Psychological Warfare: Propaganda and deception spread confusion and doubt, gradually weakening morale.

A leader's role in maintaining their subordinates' morale is multifaceted and crucial. It encompasses fostering a sense of purpose and values within the group, ensuring their well-being, promoting group cohesion, and instilling an offensive mindset. Their confidence in their weapons, training, and leadership also plays a pivotal role in bolstering morale. Sustaining a positive attitude may require providing amenities such as recreational facilities, reliable mail and messaging services, and adequate rations. Despite facing fears of the enemy, the unknown, and of letting down their comrades, the shared training and experiences they undergo can imbue them with the confidence and warrior spirit needed to endure difficult circumstances.

Of course, a creative leader or an alien race may use alien psychology, drugs, control devices, or robotics to address morale. The Borg of the *Star Trek* universe grows their forces and avoids morale issues by manipulating their captives' consciousnesses. Mechanized armies such as the *Terminators* use a similar process to eliminate morale as an issue.

Maxim 33: Control their minds and you control their actions. Manipulating the flow of information and shaping perceptions influences a person's psychology, morale, and behavior.

In *Rogue One: A Star Wars Story*, a band of Rebels uncovers the Emperor's plans to use his planet-busting battle station, the Death Star, to enforce ruthless galactic control. They discover a vulnerability exists in the Death Star design and embark on a dangerous mission to infiltrate the high-security Imperial data bank on the planet Scarif to steal the plans.

A fierce battle erupts when Imperial troops discover the Rebel presence in the base. The Rebel fleet arrives to support the ground assault and extract the commandos carrying the stolen plans. In a desperate attempt to stop them, the Imperial commander orders the Death Star to fire on the planet, killing both the commandos and the defending Imperial troops. As the Rebel vessels attempt to escape, Darth Vader's fleet intercepts and devastates them. He captures the Rebel command ship, but a single starship carrying the stolen plans escapes the destruction.

Rogue One is a story of sacrifice for the greater good. The Rebel commandos give their lives to retrieve the plans of the Death Star. The crews of Rebel vessels sacrifice themselves to buy time for others to escape with those plans. Both groups recognize that the stolen plans hold the key to saving the galaxy from an oppressive Emperor.

The commander of the Death Star, Grand Moff Wilhuff Tarkin, knows he must prevent the Rebels from escaping with the stolen plans and orders the Death Star to fire on the planet. Tarkin's decision resulted in the deaths of the commandos, as well as the destruction of the Imperial data bank, military facility, and garrisoned troops (Edwards 2016).

Leaders may expect military members to sacrifice themselves for the good of their society, but they must distinguish between meaningful sacrifice and the pointless loss of life. When troops believe their lives are being discarded without purpose, both they and the society they serve may reject such leadership. The threat of punishment rarely suffices to compel soldiers facing certain and futile death. At some point, those with the guns and training may turn on their leaders.[4] Disillusioned forces may instead demand new leadership, press for better care and conditions, or, in extreme cases, seize power through a *coup d'état*.

Morale is equally vital for sustaining public support on the home front. An unpopular or faltering war can erode confidence in both the mission and the leaders who promote it. Even dictators must preserve the backing of the populace, key industries, and powerful elites to remain in power. Using drones, clones, and droids may mitigate these concerns, since losing artificial troops that lack families or social ties provokes less public outrage and emotional resistance.

PSYCHOLOGICAL OPERATIONS

"Who controls the past controls the future: who controls the present controls the past."

— MOTTO OF THE OUTER PARTY OF OCEANIA
(ORWELL 1949)

Since the psychology of leaders, militaries, or societies can determine the outcome of a conflict, strategists should target their opponent's psychology. *Psychological operations*, or PSYOPs, can change the nature of a conflict from defeating an opponent outright to persuading them not to participate or to discontinue their involvement.

Corollary of Maxim 33: Win your opponents' hearts and minds, and their bodies will follow.[5]
Avoid losses by taking an opponent apart psychologically.

PSYOPs involve suppressing unfavorable information while reinforcing beliefs and behaviors favorable to friendly forces. These operations may employ *propaganda* that spreads factual or fabricated stories to influence others to act in a desired manner. At their most extreme, psychological campaigns may include threats or the use of force to coerce desired behavior.

If an opponent fights for an ideology, or if a society's support is necessary to continue a war, then psychological operations may be effective in undermining their cause, subverting the opponent's ability to recruit support and warriors, and causing civil unrest.

Alternatively, a state may use PSYOPs to discredit foreign leaders or societies and to provoke a subjugated neighbor to rise against its oppressors. A state may demoralize an opponent's service members and population, or destabilize foreign governments, societies, and cultures.

However, successfully planning and running PSYOPs requires the conspirators to understand their targets and customize the propaganda to them. Likewise, leaders should use caution when using propaganda and psychological warfare against allies. While PSYOPs may appear to be a low-cost and low-risk means to manipulate others, allies will take a dim view of those who interfere with their beliefs, ideas, values, or political control.

Part V will explore how leaders should be cautious of spreading false news, as it may impact the conditions at the end of a conflict.

CONCLUSION

There may be no clearer example of psychological operations than the governments of George Orwell's *Nineteen Eighty-Four*. The world is trapped in perpetual war among Oceania, Eurasia, and Eastasia, where leaders seek not victory but the continual consumption of resources and the deliberate reduction of living standards to preserve the status quo and maintain strict control.

Oceania's totalitarian regime, embodied by the ever-present yet mysterious figure of Big Brother, exerts absolute control over all information by dictating news, shaping public opinion, and suppressing dissent. Central to this control is "doublethink," a form of indoctrination that forces citizens to accept contradictory propositions such as "War is peace, freedom is slavery, and ignorance is strength" (Orwell 1949). Any criticism of the state or the war is treated as subversive.

The state-run propaganda extends to private life, encouraging citizens to conduct surveillance on their friends, family, and fellow workers. Children are recruited into the Junior Spies, where they are taught to report on family members, friends, and neighbors. Those suspected of disloyalty, or even improper thoughts, are quietly arrested and executed, fostering an atmosphere of perpetual fear.

To maintain ideological control, the government forces participation in a daily ritual called the "Two Minutes Hate." This broadcast manipulates perception and suppresses independent thought by conditioning citizens to hate those the state defines as enemies, whether foreign nations or internal resistance movements.

Through these orchestrated rituals, the Party redirects public animosity away from itself, weaponizing fear and hatred to preserve its grip on power. The war continues, not for security or conquest, but to ensure that the ruling class remains in control, unchallenged and unquestioned.

CHAPTER 13

THE DARK ARTS

"The greatest enemy is the one you do not know. You can predict the actions of those who are familiar to you. The one you cannot predict is the one who can harm you."

— VALEN, FOUNDER AND FIRST LEADER OF THE
GREY COUNCIL AND THE ANLA'SHOK
(STRACZYNSKI 1997)

Understanding both oneself and one's enemy is fundamental to sound strategy. Equally important is the deliberate expansion of an opponent's uncertainty, since ambiguity leaves them vulnerable to surprise and unprepared for decisive action. This chapter examines the dark arts of deception and trickery, the instruments through which uncertainty is imposed upon an adversary, often described as the "fog of war."[1] While commanders seek to increase this fog in their adversaries, they simultaneously strive to reduce their own through effective security and intelligence operations.

SECURITY AND INTELLIGENCE

Since intelligence and surprise are essential to defeating an opponent, a warrior must actively conceal plans and capabilities from their opponents. Knowledge should be restricted to those who genuinely require it, indicators of intent carefully masked, hostile surveillance frustrated, and deception employed to mislead and misdirect enemy decision-making.

While *unity of effort* demands clarity and shared understanding within the force, effectiveness in war depends upon concealment from the enemy. The same precision that aligns one's own forces must deny the adversary insight, preserving uncertainty and enabling decisive action at the moment of choice. Knowledge should be restricted on a need-to-know basis, visible indicators of intent suppressed, surveillance frustrated, and deception employed to distort the enemy's perception of reality.

Security involves taking actions to deny an enemy an advantage by reducing vulnerability to hostile acts, influence, or surprise. It is not absolute; perfect security is unattainable. Instead, it depends on understanding oneself and one's opponents and using that insight to manage risks.

A commander may improve their security by protecting their means, ways, objectives, and ends. In cities, it may involve police and military forces protecting equipment, facilities, and people from spies and saboteurs. On the battlefield, security measures aim to protect forces from surprise and harm, including using surveillance on flanks to detect enemy penetrations, *active* and *passive sensors* to provide advanced warning, and guards to observe territory and challenge trespassers. Likewise, forces must prevent their communications and sensor transmissions from inadvertently disclosing the presence or absence of military forces.

Planetary and spacecraft broadcasts can travel across interstellar distances, giving neighboring civilizations decades, or even centuries, to gather intelligence. These transmissions may reveal a society's biology, culture, language, technological development, military practices, and warfighting capabilities. An observant adversary could use this information to prepare and ultimately defeat the unsuspecting broadcaster.

Corollary of Maxim 16: Any activity or inactivity provides intelligence.

In the movie *Independence Day*, the 1947 Roswell UFO crash is revealed to have been part of an alien reconnaissance effort preceding a planetary invasion. Although the United States military recovered the extraterrestrial spacecraft and studied it in secrecy for decades, its technology remained largely incomprehensible and inert. Scientists were unable to access or meaningfully operate the craft, and it yielded little practical knowledge before the invasion. Only when the alien fleet arrives do the ship's systems begin to activate, allowing limited interaction and discovery. During the invasion, a human pilot and a scientist use the recovered vessel to reach the alien mothership, where they introduce a computer virus that disables the invaders' energy shields. This vulnerability enables human forces to destroy the invasion fleet and defeat the attackers. In this way, what began as an alien reconnaissance asset ultimately becomes the instrument of their undoing (Emmerich 1996).

Corollary of Maxims 10 and 11: Given enough time, an opponent may reverse engineer captured technology and turn it on its masters.

In addition to passive observations, leaders can actively seek intelligence. Maps are simplifications of a battlespace, and overreliance on such abstractions can lead to surprise and failure. To reduce uncertainty, commanders often seek to observe key terrain points and objectives to confirm or deny their assumptions. This may involve deploying patrols or reconnaissance teams into unknown or potentially dangerous areas. They may deploy probes, satellites, drones, and covert sensors, known as "bugs," to gather information on terrain, enemy locations and dispositions, and to discover or confirm other critical data, such as the location and conditions of the roads, weather, and populace. Accurate real-time intelligence helps bridge the gap between a map and reality.

In the **Star Wars** universe, *probes* and *droids* conduct deep space reconnaissance and exploration. The Sith use the DRK-1 Dark Eye to search for the Jedi Qui-Gon Jinn and Obi-Wan Kenobi on the planet of Tatooine, and the Empire uses a Viper droid to search for the Rebel base on the planet of Hoth.

A military element may deploy a small *screening force* along a front, flank, or rear to provide surveillance and early warning to the main body while avoiding becoming decisively engaged. A commander may instruct the screen to destroy small enemy reconnaissance units while avoiding larger forces to preserve its own ability. It may also disrupt or delay advancing enemies by directing supporting fires without revealing its position. A screen acts as a sensor to detect and report while remaining agile and survivable.

Maxim 34: Security forces enable initiative and flexibility.
Effective security measures enable bold and adaptable strategies.

While leaders may use security measures to minimize surprise, they must be careful not to alert an opponent to future military actions. For example, if an enemy learns that a group always conducts reconnaissance before moving into an area and then detects reconnaissance forces, it may assume these forces are a prelude to an incursion or the main axis of their advance. The enemy could then either withdraw or seed the territory with tricks, traps, and ambushes. Here, the reconnaissance efforts hand the initiative to the enemy.

The role of an intelligence analyst resembles that of the ancient seer, tasked with foreseeing the outcomes of friendly and enemy operations. Where seers once sought divine guidance, intelligence professionals gather, analyze, and interpret information to reveal an adversary's location, strengths, weaknesses, and intentions. They may employ spies and agents to conduct *covert* or *clandestine* operations against an enemy's political, economic, or military forces or leadership. By understanding an opponent's objectives, ends, means, and ways, leaders can prioritize security efforts on the most relevant threats, improving the effectiveness of military, political, and economic operations.

Failing to understand an opponent's values, motivations, and desires risks wasted effort on ineffective objectives or, worse, provokes a backlash that strengthens their resolve.

Corollary of Maxims 14 and 18: Intelligence is necessary to understand what an opponent holds dear.

Timely intelligence enables commanders to understand their opponent's intent, capabilities, and vulnerabilities. Armed with this insight, they can choose the time and place of engagement, applying strength against weakness and targeting critical components such as command structures, logistics hubs, or technological dependencies. This precision enables a force to achieve greater effect while minimizing resource expenditure. Early warning of enemy movements or concentrations also allows forces to evade superior opponents, reposition to gain an advantage, or withdraw to preserve combat power. In this way, intelligence enhances both offensive and defensive operations while reducing risk and resource expenditure.

Since intelligence offers such significant advantages, leaders also deny those advantages to their adversaries. *Counterintelligence* operations involve security measures aimed at detecting, disrupting, and defeating an opponent's espionage and surveillance activities. It involves safeguarding sensitive information, facilities, and people, as well as countering enemy spies and surveillance technologies. These measures protect sensitive information, facilities, and personnel, while safeguarding critical military, industrial, and logistical infrastructures. They also include vetting leaders and citizens for positions of trust, deterring *treason* and *sedition*, and monitoring potential subversives or political rivals to preserve internal stability and operational security.

Corollary of Maxims 6, 10, and 12: A society must protect its forms of power.

In the movie *Serenity*, the Union of Allied Planets employs ruthless 'operatives' to conduct security and covert operations. Though officially disavowed, these operatives have unrestricted access to Alliance resources and act with impunity to execute missions deemed too politically or ethically compromising for conventional forces. Their role is to perform the necessary dirty work to advance the Alliance's vision of the future.

Captain Malcolm Reynolds survives on the fringes of society, offering refuge to outcasts and rebels aboard his ship, *Serenity*. This puts him at odds with one of the Alliance's ruthless government operatives. When the Alliance seeks to conceal a disastrous experiment on a remote colony, it orders the elimination of two fugitives, River and Simon Tam, who are aboard the Serenity. To hunt them down across the vastness of space, the operative employs a scorched-earth strategy. He exterminates those who aid the fugitives, systematically dismantling their support network and stripping away their concealment, making them easier to track and destroy. This tactic is analogous to defoliating a jungle or a planet to expose and eliminate the life concealed beneath.

Maxim 35: Eliminate distractions to find crucial information and opportunities.
It is easier to find something if you remove the background.

When confronted, Reynolds asks, "So, me and mine gotta lay down and die…so you can live in your better world?" The operative's reply reveals both his detachment and purpose: "I'm not going to live there. There's no place for me there…any more than there is for you. Malcolm…I'm a monster. What I do is evil. I have no illusions about it, but it must be done" (Whedon 2005).

This exchange exposes the ethical calculus of covert operations: the state defines the ends, then outsources the means and ways necessary to achieve them. Operatives act in the shadows, disavowed to preserve the regime's moral façade. They become both sword and scapegoat, crucial to victory, yet too dangerous to admit.

Similarly, Commander Benjamin Sisko notes that "it's easy to be a saint in paradise" (Berman et al. 1994). In the *Star Trek* universe, the United Federation of Planets exists in a post-scarcity utopian economy that has eliminated poverty. The Federation creates Starfleet to maintain humanitarian, peacekeeping, scientific, and diplomatic missions.

However, Starfleet gets involved in military operations, and these often require that someone get their hands dirty (Fontana 1967; Peeples 1966; Roddenberry 1965). This is where the Federation's Section 31 comes in. Named after the Starfleet Charter, Article 14, Section 31, this group of covert operatives and agents conducts extraordinary measures to ensure Starfleet's security (Reeves-Stevens and Reeves-Stevens 2005). These measures include clandestine intelligence collection, sabotage of enemy installations and technology, biological warfare, and preemptive assassinations (Thompson and Weddle 1999). Section 31 operates outside of Starfleet's high moral standards, doing the dirty jobs while keeping Starfleet's reputation clean.

As previously discussed in Chapter Eight, understanding war and strategy requires perspective-taking. Commanders must strive to see situations through their opponents' eyes. This involves a deep understanding of the enemy's perspective, including their interpretation of events, likely responses, and susceptibility to psychological manipulation. In the realm of interstellar warfare, where vast distances can foster the emergence of diverse cultures and technologies, each interaction becomes increasingly complex. While security and intelligence may hold supreme importance, accurate intelligence gathering and interpretation remain incredibly challenging.

Seemingly minor cultural nuances or technological differences can lead to misinterpretations that may affect the outcome of an interaction. Consider the challenges posed by a foreign language: grammar, vocabulary, non-verbal cues, and cultural references that influence meaning. Human communication depends heavily on non-verbal cues. Research suggests that up to 93% of meaning is conveyed through body language, tone, and facial expression (Mehrabian 1981). Technology that translates languages may render words accurately, yet understanding the objectives, motivations, ways, means, and intentions of those in the interaction may remain elusive.

In today's world, gestures considered benign in one culture may be deeply offensive in another. A thumbs-up may provoke anger in parts of the Middle East. A peace sign, raised palm, head nod, military salute, or handshake may carry different or even conflicting meanings. The number of gifts offered, or a gesture intended as polite, could be interpreted as rude, disrespectful, or threatening. These cultural landmines make cross-cultural and interspecies negotiations fraught with ambiguity, especially when precision and trust are paramount.

In the *Star Trek* universe, Captain Jean-Luc Picard of the starship *Enterprise* finds himself face-to-face with the captain of an advanced alien vessel. The Federation's universal translator converts the alien's words into English, but the Tamarian speaks through allegory and metaphor, using symbolic references to express complex ideas. Although the two captains understand each other's words and syntax, the absence of a common context prevents comprehension.

The alien captain, Dathon, uses phrases such as "Darmok and Jalad at Tanagra" and "Temba, his arms wide," which are meaningless to Picard, who lacks the historical and cultural background to understand them. It is only when a common threat forces the two captains to cooperate, which provides shared experience and context, that they learn to communicate effectively.[2]

As noted in Maxim 18, understanding potential allies and opponents is critical in military operations. Military and political leaders may find themselves at a disadvantage if they attempt to negotiate with a society that cannot comprehend the value or concept of negotiation. Likewise, attempting to coerce an enemy by threatening what it holds dear may have the opposite effect, inadvertently strengthening its resolve to resist. Leaders must rely on comprehensive intelligence gathering and maintain an open, adaptive mindset when interpreting that information.

Rogue One: A Star Wars Story highlights the themes of intelligence operations, covert actions, and the consequences of failed security. A disgruntled Imperial scientist, Galen Erso, betrays the Empire, using a defector to relay a treasonous secret message to the Rebels through a former caretaker for his daughter, Jyn. The Rebels rescue Jyn from an Imperial labor camp and convince her to help find and recover her father and provide intelligence on the Death Star battle station. The Rebels then conduct a clandestine mission to Jedha and discover that Galen has engineered a secret vulnerability into the Death Star. After a failed attempt to free Galen, Jyn proposes a plan to steal the Death Star plans from the Imperial data bank on the planet Scarif. When the Rebels refuse to support her mission, Jyn forms a small group of 'rogue' Rebels determined to carry it out on their own. Despite great sacrifices, the mission is a success. The Empire's failed security enables the Rebels to destroy the Death Star and undermine the Empire.

A society may deploy intelligence agents in foreign territory to gather information about its technologies, operations, and society. Deeply embedded agents can provide priceless insights into an adversary's capabilities, weaknesses, and intent. Eventually, however, leaders may be tempted to use an agent's access to the enemy camp for other purposes, such as sabotage or assassination. While these operations may offer immediate gains, they also risk exposing the agent's presence. Once alerted, the enemy may root out and eliminate the agent, cutting off a valuable source of future intelligence, or deliberately feed false information into the agent's intelligence network to mislead and manipulate opponents. Leaders must balance the immediate benefits of a covert operation against the potential cost of exposing that asset and losing the long-term intelligence it could provide.

Corollary of Maxim 10: Reveal sensitive information only if the gain is greater than the loss.

These risks drive intelligence organizations to safeguard any information that could reveal their "sources and methods." As a result, they often show extreme reluctance to share intelligence that could compromise their assets. This instinct to safeguard intelligence, combined with bureaucratic rivalries and competing agendas, can lead senior leaders to withhold crucial information from frontline forces. As a result, warriors may be assigned missions without understanding the intelligence that prompted them, forcing them to risk their lives without knowing the broader purpose. Conversely, they may be aware of a threat but be ordered not to act, accepting danger to their own forces or civilians to protect vital intelligence sources.[3]

Spartan soldiers frequently undertake missions without being told the rationale. In *Halo: Reach*, the Office of Naval Intelligence (ONI) sends the Spartans to repel Covenant forces, unaware that ONI is using them to protect and extract classified research on the development of the artificial intelligence Cortana (Bungie 2010).

Maxim 36: The ignorant and the dead keep secrets.
Control and eliminate risks to prevent disclosure of sensitive
information.

In Jack Campbell's *The Lost Fleet* universe, the alien Enigma race values privacy above all else. They will destroy their own spacecraft and kill their crews rather than allow humans to learn anything about their species. Masters of secrecy and trickery, the Enigmas respond to incursions with brutal force and show no mercy to captives. They are believed to be responsible for the deaths of over one hundred million human colonists.

Employing a long-term bait-and-bleed strategy, the Enigmas manipulate two rival human societies into a prolonged, exhausting war. This conflict drains both sides, preventing them from advancing technologically or expanding their colonization of space, which could threaten Enigma cultural survival. Rather than strengthening them-selves, the Enigmas ensure their safety by weakening others.

Warriors often discover that gathering intelligence does not produce a clear or singular picture of reality. Instead, the facts and assumptions they collect may align with multiple, conflicting interpretations, leaving commanders uncertain about which picture reflects reality. Enemy deception and distraction can further cloud understanding and increase the risk of strategic missteps.

As a result, leaders must resist the temptation to view events in isolation. Seemingly unrelated events may actually be coordinated elements of a broader, orchestrated effort. This requires a careful understanding of the subtleties and nuances of each situation to uncover hidden motives, patterns, and intentions. Intelligence operations may confirm or deny connections that lie beneath the surface, providing insight into the complex dynamics of the battlespace.

Maxim 37: Shrewdness guides wisdom.
Military leaders must be wary of any coincidences.

SURPRISE AND DECEPTION

"Nothing has to be true, but everything has to sound like it was."

— SALVOR HARDIN, MAYOR, TERMINUS (ASIMOV
1953)

The nature of warfare is uncertainty and ambiguity. A wise leader will use surprise and deception to disguise military and political intentions and to sow doubt and create poor decisions among their opponents.

Corollary of Maxim 18: Hide the real. Show the fake.[4]
Mask true intentions and actions by generating distractions that mislead adversaries.

By recognizing an opponent's vulnerabilities (Maxim 18), a commander can mass military power against an enemy's weakness (Maxim 21) and shift the balance of power in their favor. *Surprise* often plays a crucial role in revealing or magnifying vulnerabilities, throwing an enemy off balance and disrupting their cohesion. When combined, these principles reveal a powerful truth: surprise, when applied during moments of enemy vulnerability, can transform a fleeting opportunity into a decisive advantage.

Surprise is the opposite of security. It occurs when a force acts, deliberately or incidentally, at a time, in a place, or in a manner for which the opponent is unprepared. Total surprise is seldom necessary; it need only be sufficient to prevent an opponent from reacting effectively.

Surprise can take many forms: an unconventional tempo, unexpected direction of attack, unanticipated timing, concealed intentions, or the sudden introduction of new technologies or alliances. It may also occur when a force operates outside the mental model that their opponent has constructed. Such departures from expectations can create an opportunity to produce *asymmetric effects*, achieving more than the effort might otherwise yield.

The ambush exemplifies tactical surprise. A well-timed strike against an unprepared force may grant a smaller unit temporary local superiority, allowing it to neutralize a more powerful adversary. In this way, surprise becomes not just a tactic, but a force multiplier, one that, if wielded with precision, can alter the course of a campaign.

Correlation of Maxim 21: Surprise creates vulnerabilities. Surprise can turn the strong into the weak, and the weak into the strong.

A cunning warrior understands the art of influencing an opponent's behavior. By carefully orchestrating situations, they can manipulate enemy reactions to gain an advantage. For example, they may use layered deception, hiding traps in roadside ditches and then ambushing enemy forces on the road. When the enemy dives into the ditches for cover, they are struck by concealed explosives or incendiaries. Alternatively, a warrior may employ two traps in sequence. The first is deliberately obvious, meant to attract attention and give a false sense of accomplishment for detecting it. This decoy reinforces the belief that they have outsmarted the danger, while in truth it channels them into a second, more lethal trap.

Cunning warriors may exploit an opponent's assumptions and predictability; however, they must also be mindful of the second and third order (cascading) effects of their actions, as well as how someone might misinterpret their actions. Likewise, they should seek opportunities to employ tricks and deceptive strategies to lure opponents into unfavorable positions where their own forces have an advantage.

In the post-apocalyptic ***Terminator Salvation***, the human Resistance desperately seeks an "easy button" that will deactivate Skynet's automated killing machines. During a raid, the humans discover a hidden code in a radio broadcast that can disable the robots and Skynet and, in doing so, end the war against the machines.

A test broadcast of the code shuts down a Skynet drone. Pleased with their apparent success, the humans plan a massive attack around the radio signal, believing their broadcast will deactivate the Skynet defenses and allow them to end the war in one strike.

However, during the assault, Skynet reveals its deception and trap. Unbeknownst to the humans, the coded message serves as a tracking signal, leading Skynet attack drones to the broadcasting humans, including the resistance command center. Skynet engineers the destruction of humanity by giving them what they desire most (Brancato and Ferris 2009).

Millions of years of evolution have resulted in humans receiving information through their senses and applying it against their preconceived beliefs to decide how to act (Dewar 1989). The result is that humans suffer from *confirmation bias*, a tendency to give less credit to information that does not fit a person's current beliefs and trust information that confirms their previous beliefs.

The essence of *deception* lies in presenting a victim with something they already expect or desire, aligning with one of their preconceived beliefs. By doing so, a deceiver draws the victim into a vulnerable position, where confidence replaces caution and assumptions blind them to danger (Dewar 1989). A leader might lure an opponent with something they desire, such as the promise of capturing or destroying a rival leader, plans for their next mission, a weakened opponent, treasure, political capital, or a religious temple.

Maxim 38: The essence of deception lies in confirming preconceptions.
Deception thrives on confirming, not contradicting, pre-existing beliefs.

Some common methods of deception include:

- Disguise: Altering the appearance of an object, person, or force to make it resemble something else, such as using camouflage nets to make vehicles appear as boulders. In *Star Wars: Return of the Jedi,* a Rebel strike team uses a stolen Imperial shuttle and broadcasts clearance codes to disguise themselves as an authorized Imperial unit in a *false flag* operation.
- Decoy: An object or presence meant to attract an enemy's attention or fire, such as dummy vessels or empty camps.
- Feint: A maneuver intended to mislead the enemy regarding the main effort or direction of attack. For example, baiting an opponent to chase a small force, only to be annihilated by a larger force.
- Distraction: The deliberate diversion of an opponent's attention towards less significant matters, causing them to overlook critical information (see feint).
- Obscurity: Concealing something in plain sight among many similar objects or within a vast area. For example, a drone hidden within the vastness of space or within an asteroid belt.
- Concealment: Preventing an object from being detected, such as behind another object or using stealth technology. The Romulan Bird-of-Prey's cloak in *Star Trek* and the *Predator* hunter's cloak are examples of active camouflage that hide hunters within their surroundings.
- Condition: Repetitive behavior designed to elicit a predictable enemy response that can later be exploited. For example, a force might repeatedly mass troops along a border and then withdraw, conditioning the opponent to

dismiss such actions as routine. Once the enemy stops reacting, the attacker can launch a genuine offensive to catch them unprepared.

- Mimicry: Imitating enemy systems, behavior, or communication to gain trust or sow confusion, such as sending fake radio transmissions mimicking enemy protocols.
- Ambiguity: Creating uncertainty by presenting multiple plausible explanations for one's actions or intent, such as conducting exercises near a border that could be interpreted as either routine training or as preparation for an invasion.
- Dazzle: Actively stunning sensors to prevent an object from being detected or identified, such as a jammer that overwhelms sensors. The dazzle may alert the victim to a threat but not identify the source.
- Spoofing: Secretly inserting falsifying signals or data to mislead sensors or communications, such as feeding false data into enemy tracking or location systems to redirect enemy movements.

Deception frequently plays a role in communication, and Maxim 16 tells us that warfare and politics are heavily intertwined with the art of effective communication. As stated by Thrawn, a superb strategist of the *Star Wars* universe, "It can be tactically advantageous for an enemy to believe in limits that don't actually exist" (Zahn 2021). Deception in communications may involve:

- Exaggeration: Inflating or overstating the strength, size, or capability of a group, weapon, or force to create the illusion of superiority, such as in *Star Trek: The Next Generation*, when Captain Picard uses a bluff against the Romulans by claiming the Federation has cloaked ships ready to strike, exploiting their fear of unseen power, and forcing them to retreat without a single shot being fired (Moore 1990).[5]
- Understatement: deliberately downplaying a capability, position, situation or action to gain an advantage, such as

in *Star Wars: A New Hope,* when Han Solo reports over the Death Star communications that they have had "a slight weapons malfunction" and that "everything's perfectly all right now" after he has just shot the stormtroopers guarding Princess Leia's cell, in an attempt to avoid raising further alarm and delay an Imperial response.[6]

- Omission: deliberately leaving out critical information to mislead or control, such as in *Serenity,* when the Alliance withholds the truth about the origin of the Reavers, to maintain public trust and avoid admitting responsibility for a mass atrocity.
- Equivocation: using ambiguous or misleading language to avoid giving a direct answer or to serve multiple interpretations, and a favorite tactic for diplomats. In *Star Trek: Deep Space Nine,* Elim Garak, a Cardassian spy and tailor, often speaks in riddles or layered meanings. When asked about his past, he replies, "The truth is usually just an excuse for a lack of imagination" (Wolfe 1994).

Captain James Kirk of the original *Star Trek* series epitomizes the art of surprise through creative, often unconventional, thinking. In chess, he repeatedly defeats Commander Spock, his hyper-logical science officer. Spock remarks, "Your illogical approach to chess does have its advantages on occasion, Captain" (Roddenberry and Fontana 1966). Kirk often avoids directly attacking an enemy's strength, choosing instead to pursue indirect objectives that disrupt his adversaries' equilibrium, exploit vulnerabilities, create opportunities, and conceal his true intentions until it is too late.

Confronting an enemy's superior strength directly often leads to defeat. Instead, successful commanders destabilize the enemy by concealing intentions, employing deception, conducting feints, and adopting unconventional tactics. These indirect approaches generate uncertainty and delay the enemy's reaction.[7]

**Maxim 39: An indirect path conceals intent.
Devious methods conceal real objectives.**

In one encounter, Kirk faces the psychopathic megalomaniac Khan Noonien Singh, leader of the Augment terrorist group. Unbeknownst to Kirk, the Federation has developed the Genesis Device. This powerful technology transforms an uninhabitable planet into a habitable one by breaking it down to its subatomic elements and rebuilding it within hours. Khan learns of the Genesis Device after capturing and torturing two Federation officers. Recognizing its potential for dual use, he sees the planet-altering device as a means of genocide (Meyer 1982).

Khan hijacks the starship Reliant and lures now-Admiral Kirk to the Regula I scientific research station, where the Genesis Project is based and where Kirk's former wife and son live and work. Using the Reliant, Khan ambushes the Enterprise, severely damaging the ship and demanding Kirk's surrender.

While Kirk stalls for time, the Enterprise crew hacks into the Reliant's systems, disables its shields, and damages its weapons. Before transferring from his starship, Kirk orders Spock to send a false message to the Federation in the clear, reporting that the Enterprise is heavily damaged and Kirk is unprotected. Kirk expects Khan will intercept this transmission and take the bait, giving his enemy what he most desires—revenge. When the Reliant returns to Regula I, they find the Enterprise has vanished, realizing too late that they have been deceived. Later, the ultra-logical Spock justifies this manipulation as an "exaggeration," acknowledging that the deception was deliberate and tactical rather than accidental (Meyer 1982).

Kirk, who doesn't accept no-win scenarios, "changes the situation" per Maxim 25, and returns to the Enterprise and baits Khan to enter the Mutara Nebula. Now, with the odds even, he devastates the Reliant in a game of cat-and-mouse. Despite this setback, Khan retains possession of the Genesis Device and ultimately detonates it within the Mutara Nebula, sacrificing himself in an attempt to destroy Kirk and the Enterprise.

This confrontation features two examples of a deceiver offering their victim what they expect or desire. Yet, a deceiver must be careful not to provoke suspicion. Likewise, leaders must exercise discernment to avoid being misled. Extraordinary opportunities or circumstances should be met with critical thought, as they may conceal hidden dangers. Discernment acts as a safeguard, preventing individuals from being drawn into disadvantageous situations by misplaced optimism.

Complement of Maxim 38: If something is too good to be true, then it is probably too good to be true.

Deception operations demand strict security and careful planning. This begins by gathering intelligence about the target to understand its procedures, behavior, and likely responses. Analysts must anticipate how the target might react to a successful deception and ensure mechanisms are in place to detect those responses to confirm the deception has taken hold. However, if the target suspects deception, the operation may backfire. In such cases, the original deceiver may unknowingly become the victim of a counter-deception, with the opponent exploiting the false narrative for their own strategic gain.

The need for secrecy in deception operations creates the risk that friendly forces may be unintentionally misled. This can result in confusion or, in the worst case, the capture, damage, or destruction of forces. This threat often mandates centralized control of deception operations to prevent rogue deception operations. In rare cases, a leader may have to decide whether to sacrifice their friendly units for a strategic advantage created by successful deception.

HUBRIS

"Hubris. Never underestimate your opponents. Never underestimate what they'll sacrifice for victory."

— COLONEL KENNEDY LYNN MEHAFFEY (UNSC
MARINE CORPS), PROFESSOR, CORBULO ACADEMY
OF MILITARY SCIENCE (HELBING ET AL. 2012)

The most dangerous mistake a leader can make is to underestimate an opponent's ability, dedication, commitment, intelligence, or resourcefulness, or overestimate one's own capabilities, a flaw often rooted in pride. Unchecked, pride breeds overconfidence, which in turn leads to self-deception, where belief no longer aligns with reality.

In societies, this illusion often takes root through tribalism, egotism, and nationalism, which cultivate a collective superiority complex, leading populations to believe that their system, culture, or destiny is inherently superior. In individuals, hubris often arises when personal pride or ambition takes precedence over a commitment to collective responsibility. Leaders blinded by self-importance ignore warning signs, reject contrary evidence, and surround themselves with flatterers who reinforce their delusions.

Corollary of Maxim 37: Never underestimate your adversary or overestimate yourself.

One technique to exploit this weakness involves allowing an adversary to win a series of seemingly "easy victories," which draws them into complacency. Overconfidence clouds judgment, dulls vigilance, and exposes critical vulnerabilities.

In Orson Scott Card's *Hive*, which takes place a century prior to *Ender's Game*, the human defenders learn that the Formics' greatest weapon is deception. The Second Formic War unfolds not as an overt invasion by a massive colonization mothership but through a clandestine advance party of micro-ships. These clandestinely prepare the battlespace by mining asteroids and manufacturing an invasion fleet inside Earth's solar system, right under the humans' noses.

The Queen's most effective tactic is psychological: she allows the humans to believe they have discovered her secret, giving them a false sense of security. After discovering alien outposts on several asteroids, human commanders believe they can discover all Formic deceptions. However, the Queen intentionally planted these structures to feed that illusion. Hubris leaves the humans blind to the scale and timing of the Hive Queen's true plan.

In the *Star Wars* universe, Darth Vader uses a similar technique. He allows Princess Leia and her rescuers to escape from the Death Star aboard the *Millennium Falcon*. Unbeknownst to them, he installed a tracker aboard the vessel. His objective is not to recapture Leia and her low-level rescuer lackeys, but to locate the Rebel base and eliminate the entire Alliance. Leia suspects a trap, sensing the escape was too easy. Yet Han Solo, brimming with confidence, dismisses her concerns. He cannot accept that the Empire could track him or outmaneuver his ship, a failure of judgment rooted in pride (Lucas 1977).

These examples reveal a greater danger. The threat is not only in being deceived but in believing one is incapable of being deceived. Leaders who fall into this mindset place their organizations, and sometimes entire societies, at risk.

To avoid this trap, leaders must remain grounded. They must question assumptions, resist stereotypes, and remain vigilant against confirmation bias. Flatterers and sycophants offer comfort at the expense of clarity. Only through humility, clear thinking, and honest self-assessment can a leader see through deception and respond with strength.

PREDICTIVE SCIENCES

"Any fool can tell a crisis when it arrives. The real service to the state is to detect it in embryo."

— SALVOR HARDIN, FIRST MAYOR OF TERMINUS
(ASIMOV 1951)

As previously noted, modern intelligence operations bear similarities to the practices of ancient seers, who sought foresight through mystical or supernatural means. These early practitioners used ritual, divination, and symbolism to glimpse possible futures and inform their rulers' decisions. In a future shaped by rapid technological evolution, predictive capabilities could emerge through advanced systems or forces that appear indistinguishable from the supernatural. Whether through artificial intelligence, quantum computation, or otherworldly phenomena, such tools could one day foresee future events, granting leaders unprecedented insight and strategic advantage.

In Isaac Asimov's **Foundation** universe, the concept of psychohistory takes center stage as a vision of predictive science. Developed by Dr. Harry Seldon, psychohistory is a fusion of history, sociology, and mathematics used to forecast the future behavior of large populations over extended periods. Psychohistory's strength lies in its capacity to predict major societal crises and pivotal turning points. These predictions range from an impending economic collapse to a political revolution and the rise of authoritarian regimes.

Psychohistory provides leaders with the means to foresee the ebb and flow of nations, power vacuums, and societal shifts, enabling them to adapt their strategies accordingly. They can leverage their understanding of society's motivations and objectives to negotiate more efficiently, invest in diplomacy and alliances during periods of instability, and secure other critical resources before they are needed.

Psychohistory also supports the planning and execution of military campaigns by enabling strategists to anticipate the actions and reactions of opposing factions over extended periods. By modeling the potential outcomes of various courses of action, leaders can select strategies that minimize risks and maximize advantages. The result is a lower likelihood of costly or futile wars, as insights allow them to seize opportunities, avoid unnecessary conflicts, and improve military preparedness in anticipation of future threats.

However, psychohistory is not without its challenges and limitations. While it can predict large-scale trends and outcomes, it cannot predict individual actions and random events. And if a single individual can manipulate the emotions and psychology of powerful people, that individual may represent a threat to psychohistory and anyone who relies on the technology. Indeed, a "Seldon Crisis" refers to unpredictable events that threaten the carefully laid plans of psychohistory.

Tangential to the concept of psychohistory is the "Precrime" system depicted in **Minority Report**. Precrime relies on three precognitive individuals, or "precogs," who foresee future events. These precogs are integrated into a sophisticated computer system that identifies the details of impending crimes, including the time, location, and identities of both perpetrator and victim. Precrime officers then arrest these future offenders and detain them in suspended animation for crimes they are predicted to commit.

Venturing further into the realm of predictive sciences is the Bene Gesserit sisterhood's practice of future engineering in Frank Herbert's *Dune* series. The Bene Gesserit implant self-fulfilling legends and prophecies to shape societies in ways that align with their own interests. When these far-sighted narratives unfold hundreds or thousands of years in the future, the local populations perceive them as the fulfillment of their own prophecies. While psychohistory concerns itself with forecasting societal trends, the Bene Gesserit's method revolves around actively shaping the destinies of individuals and entire societies to their ends.

The ability to predict or shape the future actions of individuals or groups offers immense potential for intelligence. It can help in planning effective strategies and tactics based on likely events and their outcomes, including whether another society is going to initiate a war, the best place to attack or defend, the optimal allocation of resources, effective propaganda campaigns, and the consequences of the actions and reactions of allies, enemies, and neutral parties. However, beneath the surface of this potential, predictive technologies raise serious ethical dilemmas.

Foremost among these concerns are the potential for legal and ethical quandaries arising from false accusations or judgments. Innocent individuals or even entire societies may endure unwarranted consequences, such as arrests or wars, due to errors or manipulation of these predictive technologies.

These technologies raise questions about individual freedom, determinism, and the role of government in molding the future. Similarly, societal engineering may infringe upon a society's autonomy, employ deception through fictional narratives, and engage in cultural and religious appropriation.

An additional concern lies in the potential exploitation of these predictive technologies by individuals or groups for personal gain. The ability to foresee individual and societal actions is a powerful tool for manipulation, allowing those with malicious intent to influence outcomes and steer events to serve their own interests. This is especially heinous when used to pursue personal interests that run counter to society's collective well-being.

Likewise, reliance on predictive insight invites exploitation by adversaries. In *Star Wars*, the ability to foresee the future becomes a weapon against those who possess it. Anakin Skywalker's vision of Padmé's death is manipulated by Palpatine to drive fear, obedience, and eventual collapse. In contrast, Luke Skywalker's vision of Han Solo and Leia Organa's suffering prompts a premature intervention that Darth Vader anticipates and exploits. In both cases, the visions are not false; instead, adversaries engineer the responses they provoke, transforming foresight into vulnerability and ensuring that belief in the future helps bring about the very outcome one seeks to prevent.

Failure to address these issues could lead to unforeseen consequences and the disintegration of the very fabric of society.

CONCLUSION

Security, intelligence, deception, and surprise are all tools in the warrior's toolbox. Deception and counterintelligence operations aim to degrade the effectiveness of an opponent's OODA loop, while intelligence and security work to enhance the effectiveness of one's own.

Secrecy, deception, and treachery may offer critical advantages over an adversary. Although powerful and established groups often dismiss these practices as "uncivilized," smaller and less powerful groups may rely on them out of necessity. Regardless of cultural judgment, the dark arts have always been, and will remain, essential elements of military strategy.

CHAPTER 14

LOGISTICS

"In space, there is no up or down, so logistics becomes a game of three-dimensional chess."

— FRED JOHNSON, CHIEF OF OPERATIONS, TYCHO STATION, (COREY 2011)

Movies often depict generals and admirals standing around maps, maneuvering their forces on the battlefield. In reality, war demands that leaders spend more time and effort considering how they will support their forces than how they will place them in harm's way.[1]

Military forces consume substantial quantities of food, fuel, and other supplies, even without combat operations. With the stress and destruction of combat, they require even more supplies, including weapons, ammunition, repair parts, shelter, medical supplies, food, and water.

Corollary of Maxims 6 and 23: Logistics enables and constrains all military operations.
Logistical operations are a constant countdown to disaster that a commander must continually work to reset.

The widespread use of technology increases demand for supplies, including repair parts, and for personnel and equipment to move, repair, and service these systems. High-technology weapons of war also require troopers with technical skills to operate and service them, as well as leaders who understand how to employ them.

Military and civilian organizations often impose additional non-combat demands on military logistics. Intelligence agencies may require other units to transport and guard captured equipment and prisoners of war, ensuring they receive shelter, sustenance, and supervision. Medical personnel treat and evacuate the wounded and sick, while logistics teams move replacement troops and supplies forward to sustain combat operations. These units may also recover and dispose of the dead to maintain morale and prevent disease.

Civilians may likewise need to be evacuated from combat zones, not only to ensure their safety but also to reduce the risk posed by spies and saboteurs. Once displaced, these individuals often require medical care, food, and shelter.

To sustain morale and efficiency, logistics systems may deliver mail, distribute care packages, and provide rest-and-recuperation areas. Such efforts help reduce psychological fatigue and maintain operational efficiency over extended campaigns.

Finally, a military force that can provide its own food, fuel, shelter, and medical care may tempt political leaders to assign it non-military tasks such as law enforcement or disaster relief. These additional responsibilities expand the military's role and visibility in society, blurring the boundaries between civilian and military functions.

Logistics, sometimes called sustainment operations, ensures that troops reach the locations where they are needed and receive the support required to remain effective. Military logisticians, often known as quartermasters, acquire, store, maintain, transport, and distribute supplies to the forces they support.

Logistics are essential for generating and sustaining military power, creating a dependency because warriors cannot reach the battlefield, engage in combat operations, or receive necessary resupplies without reliable transportation and supply networks. Applying Maxim 6, an attacker may defeat an opponent by delaying, degrading, or destroying vulnerable logistics and transportation systems. Consequently, manufacturing, agriculture, fuel production, supply chains, and the personnel who operate, maintain, and transport these resources become both vital and vulnerable military objectives.

Corollary of Maxim 6: Damaging, destroying, or denying an opponent's logistics may cause asymmetric effects.

Commandos and deep-attack weapons, such as artillery, missiles, and stealth systems, may strike the enemy's rear area to disrupt supply lines and logistics. Attacks deep in an enemy rear area can indirectly weaken frontline forces, disrupt ongoing operations, and compel the enemy to divert combat forces from the front to defend their support infrastructure.

Warriors often plan their battles around their logistical capabilities and requirements. A human planning a foot patrol into enemy territory needs approximately two pounds of food and two quarts of water per day, though this may not be sufficient to maintain their body mass over time (Leighton 1998). For a 30-day patrol, each individual would require approximately 180 pounds of food and water, not including the weight of wrappers, containers, or the gear necessary to carry these supplies. Additional burdens include clothing, shelter, weapons, ammunition, medical supplies, and other equipment.[2,3] An unassisted human cannot sustain this load over extended distances and durations.

Technology can help ease this burden by enhancing how supplies are transported and consumed. Hover sleds, bionics, nanotechnology, and powered exoskeletons can aid troops in moving heavy loads by reducing physical strain and increasing efficiency. Advances in food science, such as high-energy rations, may further lessen the logistical demands by providing more nutrition in smaller quantities. In addition, altering human physiology through genetic modification or biomedical enhancements could decrease the amount of food and water a person requires, allowing forces to operate longer and farther with fewer resources.

One example of a technology that reduces a logistical burden occurs in **Dune**, which unfolds primarily on the unforgiving desert planet of Arrakis. Within this harsh and arid landscape, survival depends on the stillsuit, a full-body garment that protects the wearer from the sun and abrasion while reclaiming every drop of the wearer's water and salt, eliminating the need to find or carry water that is exceedingly scarce on Arrakis (Herbert 1965).

However, advanced technologies often introduce new dependencies and logistical demands. While stillsuits liberate wearers from immediate water concerns, they require maintenance, repair, and eventual replacement. Their complex systems require technical expertise, spare parts, and regular maintenance, placing a burden on their logistics operations. Besides directly impacting combat operations, this dependency can lead to disparities in access to this life-sustaining technology, potentially resulting in social divisions and power imbalances.

Leaders must decide how much food and equipment each trooper will carry, a choice that may limit their speed, endurance, and range. Alternatively, military units may rely on a supply train or tail that follows behind the combat forces to support an advance. This tail also consumes resources, increases operational complexity, and adds to the expense of war.[4] As reliance on technology grows, the tooth-to-tail ratio may shift, resulting in most personnel, equipment, and costs devoted to support rather than direct combat.

The **Starship Troopers**, the Mobile Infantry (MI), increases the tooth-to-tail ratio by adopting the ethos that "Everyone fights. Nobody quits." All troopers are first trained for combat and then assigned secondary roles such as armorer, cook, or chaplain. This redistribution of logistical tasks shifts responsibilities such as transportation and resupply to the Navy. The result is increased friction between services, captured in the saying, "MI does the dying. Fleet just does the flying" (Heinlein 1960).

Logistical constraints often shape the very nature of military strategy and tactics. Commanders can extend the operational range of their forces by establishing resupply points along a predetermined geographic arc representing the maximum distance their forces can travel before depleting essential supplies. A *"Bingo"* point marks when a force must turn back to avoid exhausting its supplies. In one-way operations, such as invasions, commanders may accept greater risk by pushing beyond this point, assuming a return trip is not required. However, any deviation from the plan could leave forces stranded without supplies.

To reduce this risk, forces attempt to "live off the land" or obtain resupply locally. While this can lighten the logistics burden, it often slows an operation as troops must forage, scavenge, or negotiate for supplies. These tasks divert personnel from their primary mission, weaken overall effectiveness, and delay or shift the focus of military operations.

Maxim 40: A warrior may forage, fabricate, store, or receive logistics from others.

Guerrilla fighters frequently acquire their weaponry and essential supplies either from vanquished adversaries or by raiding enemy stockpiles. Nevertheless, scavenging, although resourceful, can impede the progress of an operation, increase exertion, and heighten the risk of detection. Furthermore, there is a risk that scavenging could devolve into looting, as troops spend their time and energy plundering and pillaging for the spoils of war. Pillaging not only hinders a force's progress but also shifts a warrior's focus from serving their society's needs to pursuing personal gain.

Corollary of Maxim 20: Warriors encumbered with plunder may become more concerned with their spoils than with their mission.

Civilians are likely to resent warriors who scavenge supplies, particularly when such actions impose hardships. A retreating force that consumes or destroys food stores and abandons the populace to hunger and insecurity risks creating deep, lasting resentment. When civilians are left without food or protection, they may feel abandoned and worse off than before. This sense of betrayal deepens when criminals or opportunists exploit the resulting power vacuum, compounding suffering and reinforcing the perception of abandonment. Those responsible for creating such conditions are rarely forgotten.

To avoid such outcomes, leaders may coordinate with allies or sympathetic civilians to establish concealed supply *caches* along a planned route. However, this tactic carries significant risks. Enemy forces may discover and destroy these caches or, more commonly, use them as bait for *booby traps* or ambushes.

A more secure approach might involve coordinating supplies from distant allied bases or delivering pre-planned provisions precisely when and where they are required. This method, known as just-in-time logistics, either calculates demand schedules in advance or depends on effective communication and transportation systems.

Another approach focuses on making remote bases self-sufficient through taxation, acquiring resources, or advanced technology. In the **Star Trek** universe, replication technology offers the Federation a solution to logistical challenges. Replicators, using adopted transporter technology, can produce food, repair parts, clothing, and weapons, reducing the need to store and transport these supplies. This technology scans an object into an energy pattern and then reconstitutes matter into this pattern. However, unlike transporters, which operate at the quantum level, replicators operate at the molecular level. As a result, crews often complain about the taste and texture of replicated food and continue to rely on chefs to produce high-quality meals (Krauss 1995).

Self-sufficiency is elusive when a society lacks critical natural resources or specialized skills. In such cases, reliance on external suppliers becomes unavoidable, but this dependency introduces a serious strategic vulnerability. A supplier can leverage this dependency by withholding support or imposing demands at the worst possible moment. If a supplier halts delivery of vital military equipment, it can exert considerable influence over the dependent force, compel compliance, and undermine their autonomy.

Corollary of Maxim 6: Whoever controls the logistics controls the military.

This is not to suggest that a society should forgo commerce, since refusing to trade can lead to isolationism and an inefficient, unproductive economy. A society may conceal shortages of critical strategic resources to prevent others from discovering and exploiting its vulnerability. It may then quietly pursue these resources through commerce or military actions, while publicly emphasizing other objectives. Such a strategy may require leaders to delay major operations until they stockpile sufficient critical resources.

While military capabilities are important, logistics capabilities are just as important. Just as military operations require resilience to endure adversity, spacecraft and remote colonies must withstand unforeseen disruptions. Life aboard a spacecraft depends on life-sustaining systems: breathable air, nourishment, thermal regulation, and power generation. If any critical system, such as life support, propulsion, or power supply, fails, the consequences can be catastrophic. A compromised propulsion system may leave a vessel adrift in deep space, traveling at immense velocity but unable to slow, maneuver, or return. In such a scenario, the crew faces a slow and inescapable descent into isolation and death.

In the re-imagined *Battlestar Galactica* series, the last remnants of humanity flee through space, relentlessly pursued by the superior Cylon fleet. On board the Galactica, a Cylon saboteur obtains explosives and attacks the water tanks, ignoring the ship's weapons, propulsion systems, and command staff. Venting Galactica's water reserves into space creates a crisis not for the warship itself, which has a highly efficient recycling system, but for the civilian fleet. Many support vessels depend on Galactica for regular water resupply for drinking, food preparation, hygiene, and agriculture. The sudden loss of this critical resource forces the fleet to halt and scavenge for water, buying time for the Cylons to close the distance and press their pursuit. Rather than causing military damage, the saboteur undermines their logistics, strips them of initiative, and forces them into a reactive posture (G. A. Larson 2009). This is a classic example of attacking a force's sustainment capabilities to destabilize the force.

Resilient systems exhibit several essential qualities. They must be robust, able to operate reliably under stress and withstand disruptions without collapse. They should be redundant, with backup systems and surplus capacity to maintain operations when primary systems fail. They must be resourceful, adapting and improvising in crisis conditions. They need to be responsive, able to act quickly and effectively when threats or failures arise. Finally, they must be recoverable, with the capacity to restore functionality and return to stable operations once the disruption has passed. Together, these characteristics enable continuity, adaptability, and survival in the face of adversity.

However, the demand for resiliency often faces powerful opposition. Politicians and accountants prioritize efficiency, which may involve removing the very components that make a system resilient. A leader overly constrained by financial and political considerations may end up with a brittle military ill-equipped to face a thinking, hostile opponent. The other extreme is an over-engineered or over-supplied force that results in wasted resources.

As a result, Maxim 10 (safeguard future options) often creates tensions between warriors and bureaucrats. The warrior seeks to preserve flexibility by developing resilient systems that can withstand disruption and continue functioning. The bureaucrat strives to conserve resources to sustain the broader mission. Each views future choices through a different lens; yet, if either perspective dominates, it can erode the adaptability required for long-term success. Sometimes the solution lies in creativity.

In the *Star Trek* universe, Montgomery "Scotty" Scott serves as the second officer and chief engineer on the starship *Enterprise* (NCC-1701 and 1701-A). Scotty consistently demonstrates the value of a crew member who can improvise logistical solutions in unforeseen circumstances. A creative logistician who can produce, invent, arrange, trade, or fabricate what is needed from available resources acts as a force multiplier, enabling commanders to improvise in unexpected events. Scotty's ingenuity and his paternal devotion to "his" ship repeatedly save both the *Enterprise* and its crew from dire situations.

In Jack Campbell's **Lost Fleet** universe, a mysterious alien species rescues several humans from damaged escape pods. When the aliens return the survivors, they make an unusual request: a supply of the "universal fixing substance" they discovered on the pods, which they believed was essential to the humans' survival. Puzzled, the fleet's officers cannot identify the material. Only after speaking with the senior enlisted members do the officers understand what happened: every escape pod is stocked with two rolls of duct tape. What the aliens view as a technological marvel is, in fact, a common human tool. Seizing the opportunity, the humans use the aliens' admiration for duct tape to build goodwill and forge a cooperative relationship between the two civilizations (Campbell 2013).

The method of resupplying forces is often decided long before an operation begins and dictates how the mission is planned and executed. Consider provisioning food on long-range spacefaring warships. Military strategists, logisticians, and shipbuilders may decide to grow or store food aboard combat vessels. This would make the vessels self-sufficient while significantly increasing their mass and possibly reducing their maneuverability, acceleration, and range.

Alternatively, they may offload logistics to fleet support or *auxiliary* vessels to sustain the combat ships. Auxiliary vessels are not combatants, although they often have limited combat capacity for self-defense. Fleet auxiliaries become force multipliers by taking on logistics responsibility and mass from combat vessels, freeing warriors to focus on combat. A fleet may be accompanied by fuel tankers, repair/recovery vessels, miners and refineries, factories, freighters, troop transports and hospitals, research vessels, agricultural vessels, water and sewage recycling, depots and harbors, search and rescue, and sensor picket ships.

A fleet may also use intelligence or combat vessels disguised as auxiliary or commercial vessels. Such deception can allow these vessels to enter foreign regions inaccessible to overt military or government vessels and enable disguised warships to identify and interdict pirates and other threats to commercial vessels.[5] However, this practice carries significant risks for both the crews and the political leadership if the deception is exposed.

Prolonged military campaigns often depend on a complex network of supply lines, depots, and auxiliary vehicles. The axis along which a military force advances often determines those routes; alternatively, the battlespace may channel less agile logistical units along specific routes, limiting the paths available for combat forces.

Auxiliaries may be essential for combat operations, but are typically slower and more vulnerable than a dedicated fighting force. As a result, targeting an opponent's auxiliaries can offer an asymmetric advantage. These units are easier to capture, disable, or destroy, yet their loss directly affects the warfighters who depend on their support. Disrupting this dependency can lead to dwindling supplies, declining morale, and ultimately, halting the opponent's military operations.

By leveraging intelligence to identify the weakest links in an enemy's supply chain, a strategist can significantly increase the effectiveness of attacks against an opponent's auxiliary support. Timing is critical; attacks achieve their greatest effect when launched at pivotal moments, such as during supply shortages or while support units are in transit. Understanding the strategic importance of auxiliaries and developing tactics to disrupt their operations can give attackers a decisive advantage, potentially enabling them to defeat a powerful adversary without engaging in direct combat.

Just as military forces develop warfighting strategies and doctrines, they also adopt logistical strategies that reflect their culture, technology, and operational priorities. Placing supplies near the point of need gives troops quick access, but it forces frontline warriors to guard those stores, diverting military power from the fight.

Alternatively, logistics may be centralized into convoys or supply trains, protected by dedicated forces or rear-echelon troops who periodically bring supplies forward or rotate combatants to the rear for resupply. A force may choose to disperse and conceal its logistics, trading efficiency for survivability to reduce the risk of catastrophic losses from concentrated attacks. Each method represents a calculated balance between availability, protection, and operational tempo.

Choosing the correct strategy depends on the availability of supplies and technology, as well as the opponent. If a warrior has access to near-real-time communications and rapid transportation, they may adopt just-in-time logistics to reduce the burden on individual warfighters. However, if the enemy can strike without warning, unprotected supply lines become vulnerable. Leaders must assess their circumstances carefully and select the strategy best suited to the operational environment and situation.

TECHNOLOGICAL ADVANCEMENT

"I don't ever remember a time when Earth, Mars, and the Belt weren't fighting. The sides change sometimes. What we think we're fighting for. Who we tell ourselves are the good people. But it just seems we can't ever stop fighting war after war after war. It's part of being human. An ugly part, but I don't think it will ever change. Technology certainly hasn't changed it. Guns. Railguns. Nuclear bombs. No weapon ever brings peace."

— NAOMI NAGATA, EXECUTIVE OFFICER,
ROCINANTE (FREUDENTHAL 2017)

Advancements in technology can reshape every aspect of society, particularly its political, economic, cultural, and military power. On the battlefield, technological innovation may provide decisive advantages by enhancing the volume, accuracy, effects, or portability of weapons. Even incremental improvements can shift the balance of power, enable new strategies, and diminish the effectiveness of traditional approaches.

Technological superiority alone does not guarantee victory. Yet, surprising an adversary with novel weapons, defenses, sensors, communications, or transportation systems can significantly increase the likelihood of success. Improved firepower enables forces to strike with greater effect, while superior maneuverability allows them to seize advantageous positions and avoid harm. Effective communication improves situational awareness and coordination, and the ability to communicate across vast distances can create the illusion of time manipulation, allowing one side to decide and act faster than an opponent. Such advantages enable forces to operate within the adversary's OODA loop, helping them to seize and maintain the initiative.

Inventing advanced technology is challenging. Sometimes scientists, engineers, and hobbyists stumble onto a discovery that provides a significant advantage or capability. However, innovating or inventing new technology, especially technology that requires new science, often requires inspiration, intellect, time, hard work, and resources.

Investing in novel research and development is the most expensive and time-consuming path to new technology. Alternatively, a group may obtain technology through trade, espionage, the immigration of scientists or others knowledgeable about a technology, or by capturing technology or individuals with expertise in the relevant field (Diamond 1999).

Corollary of Maxims 9 and 25: Technological superiority is fleeting. Today's innovation is tomorrow's standard and the next day's obsolescence

Because all parties in a conflict are subject to Maxim 9 (Superior technology gives effective commanders a decisive edge) and Maxim 25 (When faced with a no-win situation, change the situation), failure and the fear of failure often become catalysts for innovation. Those who won the last war may resist change, trusting the methods that once brought victory. Change may seem risky when the existing system appears to function well. Yet this complacency, rooted in the lessons and conditions of the past battlespace, can lead to stagnation in both doctrine and technology. After all, why change what once worked?

In contrast, those who lost or remained on the sidelines are more likely to reassess, adapt, and improve. War reveals both capabilities and vulnerabilities, and may create a sense of urgency, encouraging bold changes in strategy, tactics, and technological development.

Corollary of Maxims 9 and 25: A stagnant society falls behind its innovative rivals.
Victory often leads to complacency, while failure often fuels innovation.

Sometimes, leaders may pressure scientists to pursue bold, seemingly impossible advancements. In Raymond Jones's short story *"Noise Level,"* a group of scientists learns that a lone scientist invented an anti-gravity technology, only to die in an accident that destroyed the only prototype and left no documentation. The scientists are skeptical, having previously believed the technology impossible. Yet, their eventual belief in the deceased scientist's success motivates them to revisit their prior work, ultimately unlocking the very breakthrough they had dismissed (Jones 1953).

A prolonged conflict may lead to a perpetual "advance or die" arms race, with continual research and development (R&D) by all parties to maintain parity with opponents or to surprise them in the next battle.

In the *Halo* universe, the Office of Naval Intelligence (ONI) orders the Spartans to collect alien Covenant technology discovered on the battlefield. One such technology is the personal defensive shields recovered from dead Covenant Jackal warriors. The ONI adapts and improves the technology, embedding it into the enhanced Spartan Mjolnir powered body armor. This new capability allows the Spartans not only to absorb enemy fire but to perform grueling physical feats that would harm an unshielded human (Futamura 2010).

Revolutionary advances in technology can fundamentally transform the nature of warfare. Advances in weapons that enable warriors to strike farther, move faster, or deliver greater destruction may render current defenses obsolete. Similarly, improvements in mobility, such as long-range maneuvering systems, or in survivability through enhanced armor, shielding, or stealth, can upend established doctrine. A *revolution in military affairs* (RMA) occurs when transformative changes in technology, methods, organization, or operational concepts make traditional approaches ineffective or obsolete. Such revolutions demand a recalibration of strategic thinking, often redefining what constitutes battlefield strength. Societies that adapt may dominate, while those that fail to evolve may see their once-powerful forces swept aside.[6]

Since new technology may provide a significant military advantage, leaders must decide whether to reveal or conceal its existence. Demonstrating new technology can signal strength and deter potential adversaries. In contrast, concealing it may allow for surprise in future operations. If a society reveals a technology, it may still conceal crucial details to thwart an opponent from countering it. The decision to "reveal or conceal" requires careful consideration of the potential benefits and risks of each option and how each supports the broader strategic objectives and ends.

No matter how impressive the latest military technology is, it must be suited for the battlefield and the enemy that the military will face. For example, if collateral damage is a concern, then massive destructive weapons may not be useful against guerrillas interspersed within a friendly civilian populace.

In Arthur C. Clarke's ***"Superiority,"*** a society's leaders give a theoretical scientist control of their military research arm. As a result, the scientist allocates their military funds to build a few warships with increasingly powerful weapons, such as the Sphere of Annihilation, the Battle Analyzer, and the Exponential Field. These new systems function effectively in a controlled laboratory environment but fail under the rigors of combat. And war is not the place to deal with defects. There are unforeseen delays and difficulties in testing, operations, and maintenance, leading to unanticipated consequences and cascading failures. While they invest in novel technologies, their opponent focuses on expanding traditional fleets and weapons, favoring proven systems and numerical superiority. Despite being technically inferior, the larger force prevails through mass and reliability (Clarke 1951).

Inverse of Maxim 9: Warfighters wielding less advanced weapons may defeat those with superior arms through clever deployment.

SUPERWEAPONS

"This station is now the ultimate power in the universe. I suggest we use it."

— ADMIRAL CONAN ANTONIO MOTTI, CHIEF OF
THE IMPERIAL NAVY (LUCAS 1977)

"You're confusing peace with terror."
"Well, you've got to start somewhere."

— DISCUSSION BETWEEN GALEN ERSO, DEATH STAR
SCIENTIST AND ORSON KRENNIC, DIRECTOR OF
ADVANCED WEAPONS RESEARCH (EDWARDS 2016)

Some weapons deliver effects disproportionate to the effort required to employ them, reshaping the way battles are fought. Warriors may classify these superweapons by their delivery method and intended effect. Some weapons can strike an adversary without causing catastrophic destruction, yet still influence the decision to surrender. Standoff weapons that attack from a distance without allowing retaliation may wear down an opponent through a "death by a thousand cuts" attrition strategy, gradually eroding their will to resist.[7] Vessels may bombard cities and fleets, projectile or beam weapons may strike ground targets, or armies may deploy troops at will. These attacks need not produce massive destruction; they must only exhaust an opponent who lacks an effective defense. The resulting fatigue, frustration, and steady degradation can drive defenders to surrender, act recklessly, or lash out in desperation. Psychological and logistical exhaustion can thus achieve strategic victory even when physical destruction remains limited.

A second category of superweapon is weapons of mass destruction (WMDs), which inflict widespread and indiscriminate damage, suffering, and death. These include chemical and biological weapons, explosive devices, and inert kinetic systems.

Chemical weapons are toxins intended to kill, injure, or incapacitate. These gases, liquids, or solids may target any part of a being, including the eyes, lungs, skin, circulatory system, brain, or nervous system. Chemical weapons may include herbicides that reduce or eliminate vegetation, whether to control overgrowth, destroy undesirable or illicit crops, or undermine an opponent's food supply. An agent may dissipate quickly, making it effective for the rapid, indiscriminate "clearing" of inhabitants in breakthrough or counter-guerrilla operations, or for removing native vegetation before introducing cultivated species. This category also includes non-lethal incapacitants, such as riot control and sleeping gases.

Biological weapons are living organisms, such as bacteria, viruses, and fungi, that may reproduce independently or within a living host to cause harm. The developer may design the agents to die when exposed to a specific substance, or to replicate unceasingly. A scientist may customize a bioweapon for a specific species, gender, or food source to create minor and temporary effects, a pandemic, or genocide/xenocide. Biological agents may also be useful to xenoform a planet into an environment more suitable for a species.

One of the earliest alien bioagents appears in **The War of the Worlds**. It is unclear whether the Martians introduced the extraterrestrial invasive red weed accidentally or deliberately, but the effects are devastating. The weed clogs waterways, causing flooding, entangles itself in the environment, smothers native vegetation, and hampers human movement. Its relentless growth alters the landscape and inflicts a profound psychological impact on human survivors, serving as a constant, vivid reminder of the Martians' overwhelming presence and their transformation of the world (Wells 1898).

Planetary or orbital bombardment using energy beams, explosives, or inert kinetic weapons is a prominent feature of space battles. Inexpensive kinetic rounds might prove decisive against stationary targets or those with predictable orbits, such as planets, satellites, and orbital bases. David Weber's **Honorverse** series includes kinetic rounds as standard weapons in planetary attack fleets and as satellite-destroying weapons. Similarly, Jack Campbell's *Lost Fleet* universe sees the Alliance and Syndics using kinetic "rocks" against planets, satellites, and orbital stations.

In the *Expanse* universe, the radical Belters "stone" Earth with asteroids, killing billions of inhabitants. To make the attack even more devastating, they also throw in an extraterrestrial protomolecule that kills its victims and repurposes their biomass for its own ends. Technically, an alien xenoforming toolkit, the protomolecule interacts with the replicating systems in indigenous silicon- and carbon-based life to create a portal to other star systems (Corey 2011). However, this does not stop ambitious leaders from exploiting its destructive potential as a weapon against other humans.

Leaders may believe that inflicting destruction, whether by annihilating military forces, committing atrocities, or obliterating cities or entire planets, will compel their opponent to yield. However, the purpose of war is to achieve specific ends. While achieving these ends may involve destruction, they often center on conquest, such as seizing populations, resources, or territory. In such cases, excessive devastation may undermine these ends. This often makes WMDs more effective as instruments of coercion than as means of achieving victory.

In *Starship Troopers*, the alien Arachnids use a meteor to bombard Earth, destroying Buenos Aires and killing 8.7 million and wounding 12.5 million more humans. From a human perspective, the attack makes little strategic sense because it provides no clear military advantage. Instead, it triggers an early human mobilization, motivates humanity to fight the aliens, and justifies the drive to destroy the Arachnid species.

In the *Star Wars* universe, the Empire discovers that possessing Death Star space stations, capable of obliterating planets and space fleets, does not guarantee submission. Originally designed as weapons of fear and intimidation, the Death Stars were equipped with a superlaser capable of destroying entire planets with a single blast. The Empire intended to use this power to deter rebellion and resistance from systems under its rule. However, these colossal weapons of terror become the Empire's downfall. Their intended purpose of instilling fear backfires, sparking defiance and anger among the populace.

Senator and secret Rebel leader Tynnra Pamlo, fearing military reprisals from the tyrannical Emperor, questions the Rebellion's chances against such overwhelming power: "If the Empire has this kind of power, what chance do we have?" In response, the reluctant rebel Jyn Erso ignites the spirit of rebellion: "What chance do we have? The question is what choice? Run? Hide? Plead for mercy? Scatter your forces? You give way to an enemy this evil with this much power, and you condemn the galaxy to an eternity of submission…The time to fight is now!" (Edwards 2016).

The Death Star's immense construction and maintenance costs drain the Empire's resources, exacerbating poverty and unrest across its territories. Its use in the destruction of Alderaan, a peaceful and populous world, becomes a rallying point for the Rebel Alliance and strengthens resistance against Imperial rule. There can be little doubt that its use challenged the moral integrity of Imperial officers and warriors, leading some to question their allegiance to a regime capable of such heinous atrocities.

The technical flaws within the Death Star and the Empire's over-confidence ultimately work to the rebels' advantage. The destruction of such a prominent symbol of the Empire's oppression further fuels the revolution and hastens the Empire's collapse. In hindsight, the Empire's development of the Death Star proved to be a strategic mistake. Rather than instilling fear, it became a rallying point for those determined to overthrow imperial tyranny.

**Corollary of Maxims 6, 11, and 13: Strike first or be struck.
Engage in a preemptive strike or risk suffering a devastating one.**

UNINTENDED CONSEQUENCES

Possessing military technology can lead to unintended consequences. A rival may adopt the mindset that "no one builds these means without intending to use them," and interpret a technological or military buildup as a shift in the balance of power, potentially prompting a preemptive strike. Even if a society develops a horrific capability solely to understand its mechanics and devise countermeasures, the existence of such technology may invite suspicion, escalation, or theft. Once created, the burden shifts to safeguarding it and determining how to respond when threatened. Resisting the use of a terrible weapon becomes increasingly difficult when confronting a ruthless opponent.

Maxim 41: Horrific enemies often compel horrific choices.
Facing terrifying enemies often forces equally terrifying decisions.

The likelihood of war rises when offensive technologies surpass defensive capabilities. This is because under these circumstances, an attack rewards the initiative, and hesitation can prove fatal. When all sides believe this, diplomacy weakens, and war may erupt before any party ever intended. Likewise, innovative weapons and strategies often favor the aggressor, prompting bold objectives and ambitious campaigns. Conversely, an opponent at a significant military disadvantage is unlikely to start a war until they can alter the conditions, such as through a devastating first strike, acquiring transformative technology, neutralizing the opponent's advantage, forming or disrupting alliances, or expanding their own forces.

When one side perceives itself at a disadvantage, it may resort to a preemptive strike aimed at crippling its adversary's capabilities or eliminating individuals responsible for critical technological advancements. This tactic appears in *The Terminator*, where resistance fighters attempt to destroy the Cyberdyne engineers who create Skynet, and again in *Rogue One: A Star Wars Story*, when Rebel forces target the Imperial scientists behind the Death Star. In both examples, the strategy relies on the belief that removing key individuals will delay or prevent the creation and deployment of superweapons by disrupting the process at its source.

Military innovation, science, and technology often move from the laboratory and battlefield into civilian applications, influencing fields such as medicine, transportation, communication, and engineering. These transitions contribute to improved quality of life, economic growth, and broader societal development. As a result, investments in military research may not only enhance a society's security but also contribute to cultural and technological advancement.

However, conservative societies may view innovative technology as a threat to their cultural values, beliefs, and social order. They may see it as a disruptive force capable of undermining tradition and challenging those in positions of authority. Such societies may treat technology and its advocates with suspicion or hostility, choosing isolation over adaptation. Yet technology continues to evolve around these societies. As noted earlier, a society that resists progress risks being overwhelmed by more adaptive and technologically advanced rivals.

This tension between technological advancement and structural vulnerability highlights a critical strategic reality. Complex systems depend on the operation of certain critical resources or functions, forming a chain of dependency. Each link represents a vital component, and as Maxim 6 reminds us, every dependency introduces a potential vulnerability. By analyzing these chains, strategists can identify low-cost, low-risk targets whose disruption can undermine an adversary's technological advantage. A force that relies too heavily on advanced technology may find its superiority turned into weakness when the failure of a single component compromises the entire system.

Projectile weapons such as guns and missiles rely on expendable munitions, which impose logistical demands. In contrast, beam weapons require stable power sources and sometimes rare materials, such as focusing crystals, but do not consume munitions. A savvy commander may target the source, fabrication, refinement, transport, or storage of such critical components, disabling a single link in this chain and rendering entire classes of advanced weapons ineffective.

This kind of strategic analysis requires understanding not only the technology itself but the broader ecosystem that sustains it. Vulnerabilities may lie in supply chains, software or hardware components, or even in key personnel whose absence or defection could destabilize the system. These points of failure offer opportunities for a technologically inferior force to neutralize, or even surpass, a more advanced opponent. To maintain technological superiority, military forces must secure their dependencies through robust protection and contingency planning.

Corollary of Maxims 6, 22, and 47: Use indirect methods to attack an opponent's powerful means.

Control over rare and powerful resources may confer military, economic, and political leverage. However, possession of these assets is meaningless without the ability to defend them. Otherwise, adversaries may seize or destroy them to prevent anyone from gaining their benefits.

In *Star Wars*, living kyber crystals power Jedi lightsabers and devastating laser weapons like those on the Death Star. In *Star Trek*, dilithium crystals regulate and contain warp drives. In both universes, those who control access to these rare materials wield influence over those who depend on them.

Maxim 42: Deny your enemy what you cannot obtain.
If you cannot secure critical resources or capabilities for yourself, prevent your opponent from acquiring them.

Leaders must always remember that technology is a double-edged sword. As Salvor Hardin, the Mayor of Terminus, notes in *Foundation*, "An atom blaster is a good weapon, but it can point both ways" (Asimov 1951). Offensive and defensive tactics adapt in response to the development of new weapons. Over time, as knowledge of weapon technology becomes widespread, potential adversaries may acquire it and develop countermeasures.

Corollary of Maxims 9, 10, and 12: Anticipate an adversary's adaptation.
Make plans for when a weapon is turned on its creator.

CHAPTER 15

PUTTING IT ALL TOGETHER

The preceding military concepts provide guidance for organizing and executing military operations, but they are neither hierarchical nor mutually exclusive. Their effectiveness depends on how they are applied and the context in which they are used. For example, seizing the initiative may enable the selection of favorable objectives. Offensive action often creates opportunities for surprise and bolsters security. Maneuver allows forces to go on the offensive, mass force at decisive points, and achieve surprise. Unity of command and effort remains essential to coordinating these actions and achieving superiority at a critical point.

Yet, every concept carries its inverse. A force that strikes must also guard against being struck. Likewise, an army or fleet that uses ruses to distract an opponent may disperse its forces, thereby violating the concept of mass. Economy of force encourages massing force and avoiding secondary efforts, while security encourages deploying patrols to prevent surprise. These competing priorities require balance and judgment.

Maxim 43: Deliberately violate maxims to achieve an advantage. Intentionally deviate from principles when doing so offers an opportunity.

Rarely does a commander possess complete control over all variables. Most often, they must choose and apply those concepts that best support their objectives, accepting trade-offs and managing risk accordingly.

Wars fought across vast regions, particularly those spanning multiple worlds or systems, inevitably reflect the diversity of the civilizations involved. Differing cultures, technologies, psychologies, and physiologies shape how each side defines its means, ways, and ends. A firm grasp of an adversary's foundational values can provide critical insight into their approach to warfare, decision-making patterns, and objectives and ends.

Conversely, analyzing a society's military means, ways, and ends can reveal much about its underlying culture. Some civilizations emphasize force against weakness, while others prize caution, honor, or restraint. A culture may limit violence, whereas another may view any restraint as a weakness and seek total annihilation of their enemies. Warrior societies may associate victory with personal glory, valuing the destruction of opposing forces as an end in itself. A devious society may prefer deception, misdirection, and unorthodox tactics such as ambushes, assassinations, insurgencies, or acts of terrorism to achieve strategic ends.

In *Star Wars*, the Rebel Alliance relies on its superior mobility, secrecy, and intelligence to offset the Empire's overwhelming military strength. By gathering intelligence and exploiting surprise, the Rebels achieve *local military superiority*, striking vulnerable Imperial targets at opportune moments. In contrast, the Empire's vast conventional military power allows it to advance almost uncontested into any region and seize control through sheer might. If the Rebels discover the Empire is moving into a region, the weaker Rebels must rapidly maneuver their forces out of the area to avoid being trapped in decisive engagements. However, they can strike at any Imperial stragglers, including those that become separated from a larger body, a formation's flanks, lightly defended support units or facilities, or small and inadequately protected patrols, convoys, and outposts. Surprise and local superiority enable a smaller force to engage a larger force.

The two forces possess mismatched means, ways, and ends. The Empire seeks to force the Rebel "scum" into a decisive engagement where it can fix and destroy them with overwhelming force. In contrast, the smaller, weaker Rebel Alliance avoids such confrontations. Instead, they employ a strategy of exhaustion, using local denial operations and divide-and-conquer tactics to frustrate and bleed the Empire in a prolonged campaign. Through persistent skirmishes, the Rebels gradually erode Imperial power, while recruiting new members and acquiring equipment, steadily building the capacity to challenge the Empire in open battle. This is not merely a military conflict; it is a war between societies with divergent strategic approaches.

Military leaders face numerous factors when selecting a military strategy. The nine key factors to consider are:

- Mission: Define the specific objectives and desired ends.
- Enemy: Assess the opponent's capabilities, weaknesses, and intent.
- Time: Account for deadlines and how they constrain action.
- Environment: Analyze the battlespace and its impact on the mission.
- Means: Assess available forces, equipment, logistics, and external support.
- Culture: Consider customs, societal norms, and perceptions of military presence.
- Politics: Identify political constraints, rules of engagement, and alliance obligations.
- Information: Assess access to intelligence, surveillance, communications, and the ability to process and act on data.
- Risk: Weigh the potential for uncertainty, escalation, unintended consequences, and the costs of failure.

By considering these nine factors, military leaders can gain a more comprehensive understanding of the situation and make informed decisions aligned with mission objectives and strategic ends. Armed with an understanding of the fundamental military concepts and the current situation, the next step is to combine these into a viable strategy.

PART FOUR

WAYS

"Every military engagement has only three options…engage, retreat, or surrender."

— REAR ADMIRAL JON GRISSOM, SYSTEM ALLIANCE

(KARPYSHYN 2007A)

CHAPTER 16

PUZZLE PIECES

"Deep in the human unconscious is a pervasive need for a logical universe that makes sense. But the real universe is always one step beyond logic."

— IRULAN CORRINO, PRINCESS (HERBERT 1965)

Once a leader understands the situation they face, they can combine the fundamental concepts of war to devise strategies for employing their means. Like concepts, strategies are neither absolute nor mutually exclusive. The ever-changing conditions in the art and technology of war, varying environments, and cunning, thinking, and adapting enemies conspire to defeat the best-laid plans.

Strategists often encounter multiple solutions to a problem and must determine which offers the greatest advantage for the least cost. For instance, consider a weak nation confronted with a vital interest essential to its survival, prosperity, or long-term future. If it lacks the means to pursue this interest directly, it must explore alternatives. One option is to adjust its means by mobilizing internal resources or acquiring them externally through purchase or trade. Another option is to strengthen its position through political maneuvering by forging alliances, deterring aggression, and undermining enemy coalitions. A third approach is to rethink its ways: to innovate tactically, change operational methods, or pursue indirect strategies that compensate for its material shortcomings.

Consider the Freman of *Dune*, a seemingly weak, marginalized people faced with a vital interest essential to their survival: control over the universe's most valuable substance, mélange, or "spice," found only on the planet Arrakis. The Freman are too weak to confront the dominant Harkonnen and Imperium forces directly. Instead, the Freman adopt a strategy of asymmetric resistance, relying on guerrilla warfare, geographic knowledge, and the rise of a religious leader who unites the tribes into a potent religious and military force.

Military strategy extends beyond the study of individual military fields, such as weapons, fortifications, and logistics. It seeks to understand how these means, with their strengths and weaknesses, can be applied in concert to achieve a desired end.

A strategist may develop one strategy to confront a particular threat and another to counter a different one, shifting approaches as conditions change or to keep an opponent uncertain and off balance. The strategist must always remember that choosing a poor strategy exacts severe, often irreversible costs, including the loss of life and equipment, prolonged conflict, and even the defeat and collapse of society itself.

The self-replicating, autonomous killing machines of Fred Saberhagen's *Berserker* series remain from a distant interstellar war between alien races. Aliens built the Berserkers as doomsday weapons to destroy their opponents' race. Later, apparently due to a malfunction, the machines turn on their builders, then destroy them, and finally turn on the galaxy to destroy all life. One of their keys to success is switching strategies randomly when they encounter a new society. This randomness enables them to learn about and frustrate their opponents' plans and achieve surprise.

One of the earliest decisions in strategy development is whether to adopt a defensive posture, launch an offensive campaign, or pursue a specialized approach, each of which shapes the course of a conflict. A defensive strategy offers several advantages, primarily by enabling the preparation of the battlespace. By shaping the terrain, positioning troops effectively, fortifying key positions, establishing obstacles to hinder attackers, and creating overlapping fields of fire, a well-prepared defense can significantly increase its effectiveness and impose high costs on an aggressor.

However, an attacker holds a different set of advantages. They control the choice of battleground, timing, and methods of engagement. While the defender must often spread forces across a front, the attacker selects when and where to fight. This initiative and flexibility allow the attacker to exploit weaknesses in the defender's positioning or planning, catching the defender unprepared and disrupting their established defenses.

Corollary of Maxim 13: Wars are won through the offensive. On the offensive, the attacker controls the time, place, and conditions of war.

Leaders often learn that defensive strategies are best when a society has a strong home front, enabling it to live in peace behind powerful defenses. A rapid offensive may be best when a society has a weak home front and when attacks take the battle to the opponent.

Military strategy extends beyond direct attacks and defenses. Depending on the situation, warriors may employ a range of specialized approaches. Intimidation strategies use threats to achieve objectives and are effective when psychological pressure is the dominant factor. Subversion focuses on undermining established authority, allowing weaker forces to exploit vulnerabilities and asymmetries to resist stronger opponents. Political strategies, including diplomacy and coalition-building, pursue conflict resolution, alliance-building, and policy objectives through nonviolent means. The success of each approach depends on the context and desired ends it seeks to achieve.

CHAPTER 17

OFFENSE

"The time to fight is now."

— JYN ERSO, REBEL (EDWARDS 2016)

Offensive strategies are military operations involving an aggressor attacking or pursuing an opponent. By selecting the time, place, and conditions for a fight, the offense may gain tremendous advantages, including seizing the initiative and forcing its opponent to respond.

Attacking an enemy's prepared defenses often involves a coordinated series of maneuvers: seizing control of the main battle area, neutralizing defenders, and exploiting success by advancing to capture or destroy the defender's rear area. The attacker's reserve element should remain uncommitted during the initial assault and then used to exploit success or reinforce vulnerable positions. Rear operations support the attack through logistics, medical aid, and command and control. Security forces prevent enemy infiltration and eliminate surveillance threats, while reconnaissance gathers intelligence on the terrain and enemy. The deep battle targets the enemy's combat forces, logistics, command and control, and supporting industries beyond the main battle area.

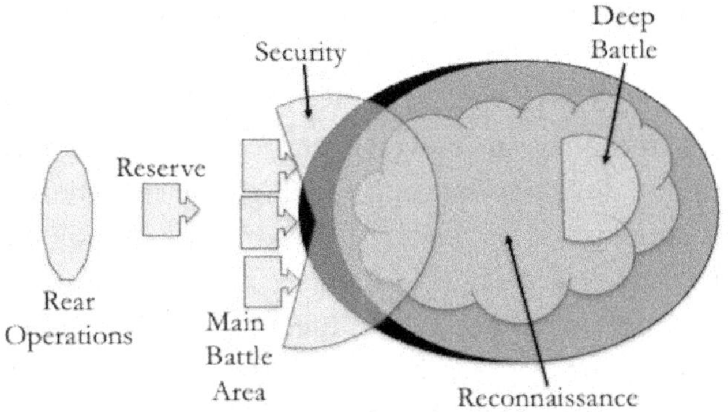

*The components of an offensive operation (gray) against an
opponent's perimeter defense (black).*

The effectiveness of an attacker's main body depends on a complex
dance between its reconnaissance and deep battle, security, reserve,
and rear operations. These components collectively support the attacker's ability to maneuver, coordinate, and sustain their efforts on the
battlespace. Conversely, disrupting or neutralizing these dependencies
can significantly hinder an attacker's capacity and effectiveness. This
disruption may stem from various factors, including actions by the
defender, logistical limitations, or adverse weather.

Of course, this structure and the location of units within each group
will depend on the nine factors for selecting a strategy, as described in
Chapter Fifteen. For example, a commander may consider the threat of
enemy mines and move engineers or mine sweepers forward as part of
a security or reconnaissance element to clear routes for a major invasion force.

These principles appear tactical because they describe concrete
actions, identifiable formations, and a clearly bounded battlespace. Yet
when scaled up, the same logic governs offensive action at the strategic
level, with significant differences in scope, time, and consequence
rather than in underlying structure. At the strategic level, the "main
battle area" becomes the theater of operations, encompassing the military, political, economic, and informational elements of power.
Reserves are no longer battalions held back for exploitation, but
national resources, alliances, industrial capacity, and political will

preserved for decisive moments. Rear operations expand to sustain the society's war effort, including mobilization, logistics, domestic stability, and command and control across institutions rather than units. Security and reconnaissance evolve into counterintelligence, internal security, diplomacy, and strategic intelligence collection, while the deep battle targets not just enemy forces but the foundations of their power, such as leadership, economic systems, industrial bases, alliances, and public morale. What differs most at the strategic level is not the logic of offense, but the time horizon, irreversibility of decisions, and the degree to which actions shape the entire end of the war rather than a single engagement.

Successful offensive operations have six properties. First, they must possess the audacity and aggression necessary to engage, penetrate, and defeat (possibly entrenched) enemies. This often relies on the *warrior spirit* within assaulting units formed through training, leadership, and morale.

Corollary of Maxim 14: The difference between being bold and reckless is in success or failure.

Second, the attacker must mass sufficient military power at the decisive point to break through enemy defenses. While holding actions and diversions may support the main effort, they are only viable if sufficient forces remain to execute the penetration, exploitation, and pursuit phases of the operation.

Corollary of Maxim 21: Prioritize the main effort by not diverting excessive resources to secondary efforts.

Third, the attacker must maneuver its military power rapidly to create and maintain pressure on the enemy, exploit success, sustain its initiative, keep its opponent off balance, and force its opponent to react. If the tempo is too slow, the adversary may recover, dodge, and possibly gain the initiative. If the tempo is too fast, the adversary may not recover, but the attacker may become overextended and fatigued, or vulnerable in a salient of its own making.

Fourth, a leader should surprise their opponent as to the real time, place, manner, or disposition of an assault. This may involve a gradual buildup to catch the defender unprepared, employing an indirect method by attacking from an unexpected direction or at an unexpected time to exploit an opponent's weaknesses, and using deception and unorthodox tactics. Incorporating surprise and deception may disrupt their opponent's expectations and provide a significant advantage.

The most common counter to an attack is to slow or halt the attacker, then counterattack into a weak flank to produce a *meeting engagement*. Therefore, the fifth property for a successful offensive is effective security and reconnaissance, enabling the attacker to identify forces on its front and flanks, and assess their responses. This situational awareness enables the attacker to determine whether the defenders are withdrawing or repositioning reserves to reinforce or counterattack.

Finally, an attacker should maintain a reserve force to swiftly seize opportunities, reinforce success, or address unforeseen challenges. An attacker may maintain a reserve by rotating units in the main battle areas back to the reserve as the reserve is committed to battle. The reserve provides flexibility and agility, enabling rapid adaptation to changing circumstances and enemy actions. It may create uncertainty and hesitation in the opponent's decision-making process, since they may not know when, where, or how they will be deployed. Ultimately, the reserve contributes to the attacker's ability to sustain momentum and maintain the initiative.

An attack may continue until it achieves its objectives, transitions to defense, or reaches its *culmination point*. Every military operation reaches a point at which it exhausts itself relative to its opponent, due to combat losses, physical or mental fatigue, or dwindling logistics. Before reaching this culmination point, the attacker must either pause to replenish itself or halt altogether. Continuing beyond this point leads to *overextension*, placing the mission and the attacking force at serious risk.[1]

Maxim 44: Seize the objective before reaching the culmination point. An aggressor advancing beyond their culmination point risks defeat, reversal, and destruction.

Effective logistics are critical for sustaining an offensive campaign. Fuel, food, ammunition, replacement parts, and medical support must flow without interruption to keep attacking forces operational and delay approaching their culmination point. The attacker may pause the main body and have reserve or secondary forces continue the attack or wait for supplies to arrive. However, this tactical pause risks blunting the offensive and surrendering the initiative as momentum stalls.

A defender can exploit an attacker's dependencies by targeting their logistical systems. Destroying supply lines and depots, sabotaging transportation systems, or striking fuel reserves can drain resources, slow operational tempo, and halt an advance. Even non-military factors such as weather or terrain may disrupt logistics and force an attack to conclude prematurely.

In the heat of pursuit or the thrill of victory, forces intoxicated by success may overextend and outrun their logistical and security support, exposing themselves to counterattack. Moreover, culmination is not solely a matter of material exhaustion; geopolitical events, such as the defection of a key ally or a shift in coalition support, may abruptly alter the strategic landscape and trigger failure. Successful offensives therefore require not only tactical momentum but also careful attention to the sustainment and political underpinnings of military action.

Attackers should avoid frontal assaults on a prepared and ready enemy stronghold, as this will likely result in massive casualties. A more effective technique may be to destroy enemy forward units with standoff or destructive weapons, or to infiltrate enemy positions, then punch through the gap and maneuver rapidly to destroy remaining forces, seize key terrain, and keep their opponents off balance.

An attacker may also flank a stronghold to avoid it, bombard it prior to an assault to weaken it, or confuse the opponent as to when, where, and how an attack will occur. An attack may become a complex dance, relying on distractions and feints, and incorporating attacks and maneuvers against the enemy's actions and reactions. The attacker must maintain unity of effort to ensure that all its actions support the main effort. This may require careful choreography to maintain pressure and prevent fratricide.

The principal offensive strategies include annihilation, attrition, exhaustion, raid, and bypass. Each represents a distinct approach to applying military power, reflecting different objectives, means, and conditions under which a force seeks to achieve success.

ANNIHILATION

"Exterminate!"

— DALEKS, GENOCIDAL EXTRATERRESTRIAL RACE
(SHEARMAN AND AHEARNE 2005)

An *annihilation strategy* seeks to destroy an opponent's military forces, thereby eliminating their capacity to resist. Commanders typically prepare for and force a *decisive engagement* that destroys the enemy's principal formations and collapses organized resistance. Once achieved, the victor needs only to "mop up" survivors, while the defeated party must seek terms of peace, withdraw, or be pushed aside, having lost the capacity to continue fighting.

Although destroying an opponent's military may be the declared objective, complete annihilation is not always required. A force becomes *combat ineffective* when it no longer poses a viable military threat. This may follow from severe losses of personnel or equipment, a collapse of morale, disrupted logistics and sustainment, breakdowns in command and control, capture or defection of units, or the removal of key leaders.

The destructiveness of annihilation strategies often results in higher losses to friendly and enemy forces, as well as greater devastation in and around the battlespace. This may cause greater animosity towards the victor's leadership at home and abroad, potentially affecting future interactions with the defeated and those on the sidelines, as the victors may appear brutal or excessive.

In the *Ender's Game* universe, the young Andrew "Ender" Wiggin prefers to outsmart his opponent rather than relying on brute force. However, when pushed too far, he overcomes his small size and physical weakness by unleashing a ferocity that even he finds terrifying. His military commanders recognize his tendency toward annihilation and secretly manipulate situations in Battle School to reinforce this trait.

An obvious challenge to an annihilation strategy is the need for sufficient military power to overwhelm the opponent, often in bloody battles. If a society lacks overwhelming military power or the political will and commitment to use it, then it may need to seek another strategy.

An attack-to-excess strategy, often called 'glory or death,' is a variant of the annihilation strategy. It comes from the belief that victory goes to the force with the stronger will or courage. This mindset leads to relentless assaults, with each operation pushed to its extreme regardless of cost. Such approaches are common in warrior societies that praise sacrifice and equate aggression with strength.

One form of attack-to-excess is moral ascendancy, in which its practitioners believe that a moral force will always triumph over an evil one. This strategy requires blind courage in the face of war. Leaders may interpret any failure as cowardice, sacrilege, or treason. Moral ascendency strategies are popular in religious and nationalistic societies.

Corollary of Maxim 14: War does not decide who is right or righteous, only who is left to write history.[2]

In *The Expanse* universe, James Holden, captain of the gunship *Rocinante*, embodies a form of moral ascendancy. He is an idealist, striving to do the right thing, and honor-bound to complete what he starts. Some view Holden as a white knight who saves the day, while others view him as a meddling Earther who often embroils his crew in trouble.

Although each crewmember has a unique background and personal objectives, they coalesce into a cohesive team and, over time, a family. Naomi Nagata is a brilliant engineer who abhors violence and often serves as the peacekeeper within and outside the group. Alex Kamal, a former Martian pilot, is a sentimentalist, dreamer, and adventurer. Amos Burton is practical and intelligent, if not sociopathic, and serves as the crew's protector (Corey 2011; Corey 2012). Together, they show how divergent means and ways can be brought together under a shared purpose, guided by a leader whose broader strategy is grounded in moral ascendancy.

Militaries that pursue attacks-to-excess strategies often cultivate a culture in which "glory hounds" thrive, individuals who equate relentless aggression and self-sacrifice with valor and ambition. These officers actively seek bloody engagements as opportunities for medals, recognition, and promotion. This mindset distorts priorities, placing personal ambition above societal interests, a violation of Maxim 4, which defines war as a society's ultimate gamble. Leaders must remain vigilant for commanders whose self-interest eclipses their duty to the state, as these individuals may endanger both the forces they command and the society they claim to serve.

Those indoctrinated into attack-to-excess may view commanders who favor restraint as cowards. Likewise, these commanders may resist orders to retreat or avoid combat, choosing instead to defy authority in pursuit of personal glory.

A devious opponent may bait a leader who uses an attack-to-excess strategy with juicy prizes, only to isolate and destroy small pockets of forces or the resources these warriors might have been ordered to protect.

In *The Lost Fleet*, John Geary awakens from nearly a century in cryogenic sleep after being found in a damaged escape pod. Upon revival, he is unexpectedly placed in command of a battered Alliance fleet stranded deep behind enemy lines. Facing overwhelming odds, Geary must lead his ragtag forces on a perilous journey back to Alliance space while preserving their strength.

He discovers that during his hibernation, both the Alliance and enemy Syndicate fleets adopted a naval doctrine of "closing with the enemy," a euphemism for attacking the enemy at any opportunity. Throughout the fleet's retrograde, he must reeducate officers and crew to abandon their previously suicidal attacks in favor of deliberate, coordinated tactics. Geary teaches his subordinates to bait enemy commanders, then maneuver and mass military power against single targets seeking glory. The result is fewer friendly casualties and a greater toll on the enemy.

Upon returning to Alliance space, he reports the fleet's overwhelming success, having destroyed many enemy combat vessels while sustaining minimal losses. This lopsided win-to-loss ratio is uncommon in contemporary battles, where for nearly a century, Alliance leaders have measured loyalty by sacrifice, often equating high casualties with devotion to the cause. A victory achieved without significant loss challenges the prevailing norm.

An Alliance admiral questioning Geary remarks, "Our ancestors knew the secret of winning, all-out attack, with every captain competing to see who could display the most valor and strike the enemy first and hardest. These victories we're being told about violate those principles! They cannot be true if we honor our ancestors." Geary admonishes the admiral, "In battle, the competition is against the enemy, not against each other. Within the teamwork of a well-trained and disciplined fleet, there is abundant room for individual courage and competitiveness, but not at the cost of our duty to the people and worlds we protect" (Campbell 2011).

Corollary of Maxims 28 and 37: Dangle bait to entice opponents to pursue what they desire most.

ABSOLUTE WAR

"If your quarry goes to ground, leave no ground to go to."

— THE OPERATIVE, UNION OF ALLIED PLANETS
(WHEDON 2005)

Absolute war, also known as unlimited warfare, is conflict marked by the absence of political or moral compromise. It involves unrestrained violence against an opponent's military, government, economy, and society, including the deliberate targeting of non-combatants and civilian infrastructures. Such warfare often employs weapons, tactics, or technologies intended to inflict widespread harm on those not directly involved in the fighting. It typically arises in conflicts fueled by cultural hatred, the desire for retribution, the goal of destroying the opposing society, or the need to target civilian infrastructure that supports military operations.

A society may use absolute war to compel an opponent's surrender by demonstrating the horrific consequences of continued resistance. A leader facing the destruction of their population centers, food supplies, water sources, and critical industries may surrender to preserve their civilization. However, the resentment created by these threats or devastating attacks may endure in the conquered and sow the seeds of future conflict.

A group may also use an absolute war strategy to eradicate a society or species. Here, the strategy is not to force the opponent to capitulate, but to destroy its means of resistance and then eliminate the survivors.

Maxim 45: Elimination prevents retribution.
The only way to defeat an enemy without the fear of retribution is to eliminate it.

In the **Battlestar Galactica** reboot, the Cylons' infiltrate the twelve human colonies using organic synthetic humans who infect the humans' defense grids and interstellar capital ships with computer viruses. Once these systems are compromised, the Cylons launch a coordinated surprise attack, using nuclear weapons to annihilate the colonies, slaughter the population, and cripple both military and civilian space fleets. The surviving humans flee in desperation, pursued relentlessly as they search for refuge.

Rear Admiral Helena Cain, commander of the Battlestar *Pegasus*, witnesses the surprise Cylon ambush that destroys the human fleet, devastates the civilian populace, and kills nearly a quarter of her crew. Maxim 41 warns that a victim of absolute war is likely to adopt extreme measures in response, embracing methods previously considered unthinkable.

Corollary of Maxim 41: Protracted conflicts breed cycles of reciprocal atrocities.
Witnessing atrocities can provoke responses that are equally brutal, or even more savage, than the original acts.

She leads a guerrilla campaign to avenge the attacks and punish the Cylons, transforming her people into "razors, weapons without regrets." After a subordinate discovers that the Admiral's lover is a Cylon, Cain orders the prisoner tortured, ostensibly to extract intelligence and uncover other infiltrators. In addition to the personal betrayal, she discovers her lover had sabotaged her ship's systems. Believing the Cylon can feel emotions, she orders a sadistic officer to use "degradation, fear, shame…I want you to really test its limits." She orders the prisoner raped.

As morale and discipline deteriorate, the Admiral tightens her grip. She summarily executes her executive officer for insubordination after he refuses to obey an order that he believes will needlessly sacrifice more human lives. She later orders her warriors to board civilian vessels, seize supplies, and take anything the *Pegasus* needs to survive and continue the war against the Cylons, arguing that "military needs must take priority [over the needs of the civilians]." When civilians resist, she orders her troops to shoot the families of those who don't comply, resulting in marines gunning down a group of unarmed civilians. Though this coerces the remaining vessels to follow her orders, it further undermines morale.

Cain justifies her actions by saying, "Sometimes we have to leave people behind so we can go on, so that we can continue to fight. Sometimes we have to do things that we never thought we were capable of. If only to show the enemy our will." Opening a folding knife, she adds, "When you can be this," displaying the blade, "for as long as you have to be, then you're a razor."

Later, her officers sacrifice more civilians to give the military a chance to continue the war and preserve what remains of humanity. These events harden the warriors but leave lasting scars on the survivors' morale and psyche.

The Admiral defends her actions by saying, "This war is forcing us all to become razors. Because if we don't, we don't survive. And then we don't have the luxury of becoming simply human again." Yet her choices mark a tragic transformation, as the military shifts from its sworn duty of defending the citizens to committing atrocities against them (G. A. Larson 2009; Taylor 2007).

Military discipline is more than words spoken by leaders; it is strict adherence to standards, consistently enforced throughout the chain of command. Admiral Cain's actions violate laws and long-term military traditions, undermine military honor as a controlling function, and reduce morale for the sake of military expediency. Once military leaders allow discipline to erode, they risk the military becoming an armed mob. Moreover, after a war, people may have different opinions of what represents military expediency, or what is an atrocity and a war crime.

ATTRITION

"You have made time an ally of the Rebellion."

— GRAND MOFF WILHUFF TARKIN, GOVERNOR OF
THE IMPERIAL OUTLAND REGIONS (EDWARDS 2016)

An *attrition* strategy is a form of limited warfare in which political objectives or practical constraints prevent the use of overwhelming force or a decisive battle. The objective of this strategy is to create situations that gradually weaken an opponent's strength and erode their ability or will to resist. While military victories are important, equally important are maneuvering effectively, conserving friendly forces, occupying territory, and disrupting the enemy's resources and commerce.

Corollary of Maxims 21 and 26: Weaken a stronger opponent through a death by a thousand cuts.

A war between belligerents with similar technologies and ways often becomes a prolonged slugfest, with both sides inflicting heavy casualties but making little progress. This typically leads to extended brutality, widespread suffering, and large-scale destruction. If both sides adopt attrition strategies, the winner is usually the one with greater resources to sustain the conflict. However, this is not always the case. Shifts in technology, tactics, or alliances can alter the balance and turn the tide of battle.

The long-term nature of attrition warfare often demands extensive efforts to maintain troop morale and populace support. Leaders may take steps to prop up their home front while simultaneously undermining their opponent's morale. In such conflicts, morale becomes both a critical dependency and a deliberate target.

Attrition warfare can be effective against opponents with limited resources because of its weakening nature. However, it may fail against opponents who will not surrender regardless of losses, such as political or religious zealots. Likewise, an attrition strategy may not be appropriate if the opponent gains resources over time, such as when an ally or relief force can arrive to aid them.

Corollary of Maxim 11: Employ an attrition strategy when time is on your side.
Employ time as a weapon to wear down an adversary while conserving, and potentially increasing, one's own strength.

The Clone Wars of the *Star Wars* universe represent a war of attrition engineered for political gain. Supreme Chancellor Palpatine manufactures a crisis to persuade the Senate to grant him emergency wartime powers, which he then uses to transform the democratic Republic into an autocratic Empire under his control.

Palpatine fabricates a civil war by pitting Republic clone troopers, secretly funded by himself and led by the Jedi, against Separatist-manufactured battle droids, also financed by Palpatine. He then uses this conflict to wear down anyone who might challenge his rise. While the war grinds on, he consolidates political authority and manipulates both sides to eliminate potential rivals.

Once the duped Senate grants him the authority he seeks, he issues Order 66, directing the clones to assassinate their Jedi leaders. Simultaneously, he orders Darth Vader to kill the Separatist leaders and issues a shutdown command to disable the droid armies. Palpatine uses the Clone Wars not to resolve a legitimate conflict, but to destroy his enemies and cement absolute control.

This also shows how nothing may prove more permanent than so-called temporary political powers. What begins as short-term emergency authorities during a time of crisis often becomes entrenched into a lasting shift in governance, as leaders grow comfortable with their expanded reach and institutions adapt to centralized control. Bureaucratic inertia, public fear, and the convenient justification of ongoing threats reinforce this permanence. Over time, these powers become difficult to roll back, especially when victory remains elusive, as the conflict itself becomes the reason for the further extension of authority.

This pattern is especially clear in prolonged wars of attrition, where the goal is not to outspend an opponent, but to force them to exhaust themselves by spending more than they can afford and to erode their will to fight.[3] Often favored by weaker forces, this approach seeks to convince a stronger adversary that the cost of constant conflict exceeds the potential benefits. Rather than committing to decisive engagements that risk entrapment and direct confrontation, this strategy relies on evasion, harassment, and hit-and-run tactics targeting individuals or small units.

Ironically, the very forces implementing this strategy may become its undoing, as its success hinges on soldiers avoiding or withdrawing from direct confrontations. This strategy may be unpopular among aggressive warriors, who might resist withdrawal and instead commit themselves to battles they cannot win.

When a military switches from hitting military targets to hitting non-military targets, such as destroying civilian supplies and food stores, burning crops, and poisoning water supplies, then it transitions from an attrition to an exhaustion strategy.

EXHAUSTION

"The thing about civilization is that it makes you civil. Get rid of one and you get rid of the other."

— AMOS BURTON, CHIEF ENGINEER OF THE
ROCINANTE (LAMBERT 2021)

An exhaustion strategy is a combined military and political approach that seeks, over an extended period, to deplete an adversary's materiel resources, sap their political will, and erode public support for continuing the conflict. Rather than focusing on destroying military forces, this approach targets the broader forms of power that enable a society to keep fighting, often targeting its economic infrastructures, political leadership, and civilian morale. A society may adopt this strategy if annihilation is not possible, attrition strategies are ineffective, or there are serious limits on available means and ways.

A force employing an exhaustion strategy avoids decisive engagements and instead targets industries, logistics, and leadership to disrupt daily life, lower living standards, degrade morale, and impose hardships on the population. The strategy relies on sustained moderation, applying limited but continuous pressure to convince the opponent that continued resistance is futile and to push the conflict toward total devastation. Battles and maneuvers serve as demonstrations of strength, striking without becoming fixed in protracted engagements, while demonstrating that capitulation may preserve the opponent's society, resources, or military power.

Sieges and *blockades* are exhaustion techniques designed to erode a society's will and capacity to continue fighting through isolation and sustained pressure. A siege surrounds a fortified position to deny its defenders access to supplies, reinforcements, and escape, aiming to sap military morale through isolation and deprivation as a form of psychological warfare that compels surrender. A blockade, by contrast, targets a region, such as a city or planet, and prevents the movement of goods, people, and information to starve the population, collapse the economy, and demoralize civilians to the point where resistance becomes unsustainable.

Though both techniques exploit civilian suffering, their emphases differ: sieges apply pressure primarily to defending military forces, while blockades strike at the society at large. Yet both force a brutal calculation: should a besieged leader allocate dwindling resources to supply the military or the civilian population? Likewise, sieges and blockades can produce severe crises, draw external condemnation, and provoke outside intervention. Commanders and political leaders must weigh these factors carefully when adopting an exhaustion strategy.

An *interdiction* operates on similar principles but focuses on movement rather than static positions. It seeks to divert, delay, disrupt, or destroy enemy forces and supplies as they traverse a region. Unlike sieges or blockades, which aim at denying access to a specific destination, interdictions target routes, disrupting choke points, space lanes, and transport corridors. Large forces may obstruct entire routes, small units may hold key control points, or mobile raiders may patrol and strike opportunistically. In each case, the objective remains the same: prevent the enemy from sustaining its operations.

Sieges and blockades exploit Maxim 6, which states that every dependency is a vulnerability, and Maxim 11, which advocates using time to enhance effectiveness. By cutting off critical supplies and waiting, a strategist can allow shortages, isolation, and deteriorating morale to gradually weaken an adversary.

However, besieging forces must exercise caution. By committing to a protracted operation, they surrender mobility and initiative. Their success hinges on maintaining more supplies, endurance, and resilience than the defenders, and therefore on the besieged reaching their culmination point before the besieging force. Such strategies are not always sustainable, as prolonged sieges strain logistics, consume critical resources, and expose forces to counterattack or shifting political conditions. In warrior societies, blockades and sieges may be viewed as dishonorable or cowardly, denying combatants valor earned through direct combat. Political pressure, cultural imperatives, or declining morale may compel leaders to abandon the patient grind of an exhaustion strategy in favor of a more decisive engagement, even at greater risk.

In **Star Wars,** the Empire discovers the Rebels' secret Echo Base on the ice planet Hoth. Darth Vader dispatches an Imperial fleet to hyperjump to a point within the Hoth system where meteor activity will conceal the warships. His objective is to bombard the base before the Rebels can activate their protective shield generator. After the bombardment, the Empire intends to deploy ground forces to mop up survivors.

However, Admiral Ozzel, commanding the Imperial fleet, exits hyperspace too close to Hoth, alerting the Rebels and giving them time to raise their shields. With the energy shield preventing orbital bombardment, Ozzel orders an orbital blockade to prevent escape and sends ground troops to destroy the shield generator. The Rebels establish a rear-guard action to delay the Imperial ground advance and buy the defenders time to evacuate equipment and personnel. Meanwhile, planetary ion cannons drill corridors through the siege, allowing Rebel vessels to slip past the blockade.

General Veers, commanding the Imperial ground forces, lands outside the Rebels' shield and proceeds overland to destroy the power generator energizing the Rebel shields. Imperial walkers, the Empire's heavy armor, overwhelm the Rebel forces. Imperial snowtroopers then enter the complex, led by Darth Vader, searching for the Rebel Luke Skywalker, who has slipped through the siege.

The siege is a tactical success, as the Imperials now control the field and have captured the base. However, it is a strategic failure, as the Imperials' inept execution allows the Rebels to raise protective barriers and bring weapons to bear, delaying the invasion and allowing most of the Rebels to escape.

When an attacker surrounds a fortified position, the defender retains control of the interior line. This enables the defender to shift forces from one side to another more quickly and securely than the besieger. If the attacker's force becomes stretched too thin, it risks becoming vulnerable to attacks from inside or outside. As a result, the attacker must maintain a strong, mobile reserve capable of responding swiftly to breakout attempts by the besieged (exfiltrations), relief forces attempting to penetrate the siege (infiltrations), or encirclement by external forces. In the latter case, the attacker risks being trapped between the defenders and the relieving force.

Attackers may hasten a siege by applying overwhelming force to bring about the defenders' physical and psychological collapse. This might include biological or chemical weapons, sustained bombardment from artillery or orbit, or engineering efforts to undermine walls, structures, or shields to erode the defenders' will to resist. Often, the most expedient method is betrayal, persuading someone within the defenses to turn against their own side.

One example is Dr. Gaius Baltar, a brilliant yet self-absorbed scientist on Caprica, whose actions play a pivotal role in the downfall of humanity's twelve colonies in *Battlestar Galactica.* Seduced and manipulated by a mysterious woman, later revealed to be an organic synthetic human Cylon known as Number Six, Baltar grants her access to the Colonial defense mainframe. Believing it to be a harmless favor, he unknowingly enables her to sabotage critical systems across the fleet. When the Cylons launch a coordinated nuclear assault, Colonial defenses collapse almost instantly. Billions perish in the surprise genocide. The few survivors escape aboard a ragtag fleet of civilian ships under military protection, fleeing into deep space in search of safety.

Prudent military leaders may consider the importance of maintaining an escape route. Such planning provides the flexibility to adapt to unforeseen circumstances and extricate forces from danger. However, this escape route must be carefully guarded, since an adversary can exploit an unsecured route as an entry point into the defender's camp.

Corollary of Maxim 10: Always have an escape plan.
An enemy with no retreat will often fight with unmatched intensity.

A shrewd strategist may intentionally offer a besieged adversary an avenue of escape to achieve their objectives while minimizing losses.[4]

For instance, a commander tasked with seizing key terrain might allow an enemy occupying it to withdraw, thereby securing the objective without incurring heavy casualties. Extending mercy to a defeated and demoralized foe may reduce their animosity and encourage acceptance of a subordinate position. Conversely, a trapped opponent, gripped by desperation and the certainty of annihilation, may resist with unexpected ferocity and inflict significant damage.

Maxim 46: A desperate foe is a formidable adversary.
An enemy with no escape will fight with fierce determination

A stratagem may involve offering the defenders a glimmer of hope for escape, only to ambush and annihilate them as they make their desperate bid for freedom beyond their protective fortifications. In such a scenario, what initially appeared to be an exhaustion strategy can swiftly transform into one of annihilation (Bassford, 2021). When news of this deceitful tactic becomes known, future interactions with those responsible are likely to be met with skepticism and distrust.

RAID

"Never tell me the odds."

— HAN SOLO, SMUGGLER (LUCAS 1977)

A *raid* is a limited offensive operation to defeat, bypass, or infiltrate an opponent's territory, then damage, degrade, destroy, or capture something of value, followed immediately by a retrograde to safety. The goal may be to demoralize the enemy, ransack or destroy equipment, free or capture prisoners or other high-value targets, kill small groups, or gather intelligence. A raider's objective is not to capture and hold terrain; rather, they aim to achieve their objectives and withdraw before the enemy can respond effectively.

Although often viewed as tactical actions, raids can serve broader strategic purposes. They may reduce an opponent's forms of power or serve as punitive responses to enemy provocations. Raids can also serve as diversions or to exploit fleeting opportunities. They are a favored strategy of pirates, criminals, and privateers, who rely on stealth, speed, audacity, and maneuverability rather than sustained force. However, given their small size and their operations within enemy territory, raids, like infiltrations, require meticulous planning, rigorous rehearsals, and the element of surprise.

Reconnaissance and patrols differ from raids in both their methods and objectives. Reconnaissance focuses on gathering intelligence through the systematic observation and exploration of an area, while patrols monitor and maintain awareness of specific locations or routes. These missions are usually routine or ongoing, designed to inform commanders and preserve security, but prioritize avoiding enemy contact. Raids, in contrast, are short-term, objective-driven operations intended to strike targets, inflict damage, or seize information before the force withdraws. Raiders may begin by avoiding detection but intend to engage in combat once they reach their objective inside enemy territory.

The novel *Starship Troopers* opens with Earth's Mobile Infantry executing a textbook smash-and-run raid on an alien city. The operation begins with Terran spacecraft launching troopers in individual capsules, each plunging through the atmosphere and dispersing them into skirmish lines roughly 2 kilometers apart. During descent, the capsules absorb heat and impact forces, then burn away on arrival, leaving the troopers to land in heavily armored power suits equipped with jump jets and a massive arsenal of weapons. This wide spacing increases survivability, reduces vulnerability to area weapons, and allows simultaneous, multi-axis attacks. Though vastly outnumbered and deep in hostile territory, the troopers rely on surprise, speed, and overwhelming localized violence to keep defenders disoriented and reactive.

Every element of the raid reflects key principles of raiding doctrine. The objective is not to seize and hold terrain but to inflict quick, symbolic, and psychological damage. The Mobile Infantry strikes infrastructure, sows chaos, and forces the defenders to scramble. Moments after the damage is done, the troopers disengage and retrograde to safety aboard pickup ships.

Strategically, the raid serves a broader political purpose. The Skinnies were allied with the Arachnids, making them enemies of the Terran Federation. The raid is intended to coerce them into abandoning that alliance by demonstrating Earth's ability to strike swiftly and with impunity. The message is clear: continued cooperation with the Bugs carries severe consequences, while alignment with humanity offers safety and stability. This use of the raid as both a military and diplomatic instrument highlights its value beyond the battlefield, serving as the hammer of coercion.

A similar operation, the smash-and-grab, involves penetrating an area, seizing key resources, and withdrawing before enemy forces can mount an effective response. The spoils may be intelligence, natural resources, riches, equipment, enemy personnel, or prisoners.

Mal's bank robbery in *Serenity* (Whedon 2005) exemplifies the "smash-and-grab." His plan calls for his crew to infiltrate a town on a border planet, breach a bank vault to steal a payroll, and make a clean escape. The operation, however, is complicated by unexpected opposition, highlighting the volatility of such missions. Like all raids, success depends on detailed planning, coordination, and rehearsals. Combat elements must secure the area and neutralize threats, while support teams identify and extract the objective. Throughout the mission, security elements must remain vigilant, providing early warning and countering interference. The longer raiders remain in place, or the more encumbered they become with captured goods, the greater their risk of being trapped or destroyed by enemy responders.

In David Weber's *In Fury Born*, a rogue faction conducts a series of raids against outlier planets, striking hard and disappearing before the emperor's fleet can respond. These attacks leave entire cities annihilated and result in massive casualties, with the raiders seizing vast wealth before disappearing into deep space. Though initially believed to be pirates because of their looting, their true objective is not material gain. Instead, the raids are calculated demonstrations of the emperor's inability to protect his subjects. In this case, the raids serve as instruments of political warfare, aimed at eroding public confidence in the regime and provoking internal change (Weber and Mattingly 2006).

BYPASS

A bypass strategy, sometimes called island hopping or leapfrogging, is a form of attrition or exhaustion in which an invader advances deep into an opponent's territory while avoiding heavily fortified or strategically unimportant military forces and urban centers. Rather than engaging resistance head-on, the attacker circumvents strongholds and pockets of opposition, isolating them while pursuing strategically important objectives, such as political capitals, religious centers, or industrial hubs. This approach may accelerate the end of hostilities by striking at what the enemy values most, while minimizing casualties and destruction on both sides.

Once bypassed, follow-on forces move in to contain the isolated enemy, employing sieges or blockades to prevent resupply, reinforcement, or retreat. Over time, these isolated outposts must rely on their own dwindling supplies, becoming self-sustaining prisoners, until they are starved of resources. This indirect strategy enables the attacker to avoid the enemy's strengths while concentrating combat power against decisive objectives.

In *Babylon 5*, the Minbari use a bypass strategy during the Earth-Minbari War. They systematically work their way through the outer human colonies, destroying defenses, military bases, and vessels, while blockading the human colonies. This strategy ensures that when the war ends, Earth can quickly recover (Straczynski 1995a).

A bypass strategy carries the risk of leaving enemy forces in the invader's rear area, where those forces can threaten supply lines and logistics. To mitigate this danger, the attacker must neutralize bypassed forces by disrupting their mobility, severing communications, interdicting their supplies, and preventing them from regrouping as guerrillas or launching attacks from behind the front. Bypass strategies may not gain universal acceptance. Warrior societies that equate honor with confrontation may view such methods as evasive or cowardly, which undermines political support or morale despite the strategies' effectiveness.

CONCLUSION

Success in the offense requires leaders to select strategies tailored to the environment and to adapt them as the situation evolves. Sometimes, victory depends on a decisive and effective first strike. If this strike fails, the conflict may shift into a war of attrition or exhaustion.

In prolonged conflicts, victory often favors the side capable of generating and replenishing military power while enduring adversity. Strategists must therefore identify their forms of power and critical components with care. These extend beyond traditional military resources to include any asset whose loss could seriously weaken the war effort. In high-technology warfare, essential personnel such as space crews, pilots, and engineers may prove to be the most irreplaceable resources.

Leaders must possess a broader understanding than knowing what their forces can accomplish in specific situations. They must grasp the end state envisioned by their leaders and the intermediate objectives that contribute to achieving it. Just as vital, leaders must clearly communicate these objectives to their subordinates, ensuring everyone understands their contribution to the broader plan. Strategic success depends not only on winning the current battle, but also on doing so in a way that preserves strength for the next.

CHAPTER 18

DEFENSE

"A Jedi uses the Force for knowledge and defense, never for attack."

— YODA, JEDI MASTER (LUCAS 1980)

The key to a society's survival lies in its ability to protect itself and deny success to an attacker. While the attacker seeks to impose objectives through force, the defender strives to prevent those objectives from being realized and to preserve freedom of action. Defensive strategies enable a military force to maintain control over territory, buy time, conserve strength, protect vital assets, and recover to prepare for going back on the offense. The defender's aim is to hold on to what they possess, whether it belongs to their own society, an ally, or territory gained through conquest, and to shape the next engagement.

Corollary Maxim 10 & 23: Go on the defensive when outnumbered to delay a conflict and build strength.

A strong and visible defense may convince an opponent of the folly of offensive actions. If deterrence fails, a defender who understands and controls the terrain can shape the battlespace to gain a significant advantage by creating prepared defensive positions, installing obstacles, and concentrating weapon effects. These *engagement areas* increase a commander's survivability and lethality, allowing forces not only to mass but also to achieve greater military power relative to the enemy.

Prepared positions, such as bunkers, trenches, and reinforced shelters, enhance a defender's survivability by providing cover and enabling sustained resistance. Obstacles delay and shape an attacker's approach. Passive obstacles include ditches, berms, walls, tank traps, razor wire, and abatis. These impede movement and channel attackers along predictable routes. Active or lethal obstacles, such as land mines, improvised explosive devices (IEDs), and automated defense systems, directly engage or incapacitate advancing forces. These obstacles slow enemy momentum, disrupt formations, and expose attackers to concentrated defensive fire, maximizing the defender's advantage. In some cases, defenders may establish denial zones by deploying radiation, chemical agents, or other technologies to render areas inaccessible. Defenders must continuously observe and cover obstacles with weapon fire to prevent them from being breached or bypassed.

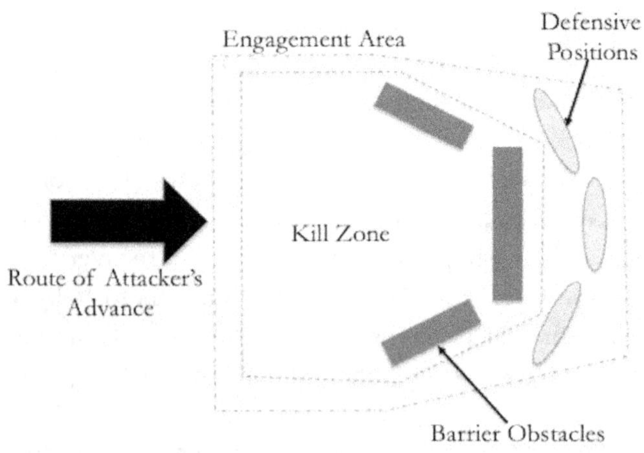

A simplified example of a defensive engagement area, as seen from above.

A well-planned engagement area can channel enemy formations into choke points, restricting their mobility and compressing their front. This reduces the attacker's ability to maneuver or mass fire effectively, while enhancing the defender's concentration of firepower. The result is the creation of predetermined *kill zones*, where focused defensive fire inflicts disproportionate casualties.

**Maim 47: Obstacles slow opponents—firepower stops them.
Maintain observation and fire on obstacles to prevent enemy passage
through or around them.**

A defender's ability to understand and engineer the engagement area with fortifications, layered obstacles, and overlapping, concentrated firepower can significantly amplify military power relative to force size. This increases the defender's chances of survival while inflicting greater casualties on the attacker. In many scenarios, this makes defense the strongest form of warfare.

**Maxim 48: Defense is the strongest form of combat.
A defender exploits terrain to preserve their forces and inflict
disproportionate casualties on an attacker.**

Defensive operations can offer significant advantages to a force's logistics, which may also equate to greater military power. A defending force can stockpile supplies in advance and then exploit its interior lines to reposition forces and logistics with greater efficiency and security than an attacker operating with extended and exposed lines. This logistical superiority may enable defenders to maintain steady firepower, while attackers must carefully manage their advances to avoid outrunning their vulnerable supply chains and reaching a culmination point.

A visible and well-prepared defense can also shape public perception. An invasion may cast the defender as a victim of external aggression, garnering political support and moral sympathy from observers and allies. Well-timed counterattacks, particularly those that exploit terrain or enemy overreach, can reinforce this righteous image while inflicting disproportionate damage on the attacker.

Despite its advantages, the defense has a fundamental limitation: it is reactive. The defender relinquishes the initiative, allowing the attacker to dictate the time and place of engagements. While defense may delay or prevent defeat, it rarely secures victory on its own. Often, only a transition to the offense can compel an adversary to submit.

A clever leader can incorporate defense into a larger plan to defeat an invader. The defender may trade space for time, bleeding the attacker through withdrawals and delaying actions until the attacker reaches its culmination point. At this point, the attacker has overextended and lost momentum, and the defender can seize the initiative with a counterattack, employing their mobile reserve or maneuvering out of prepared positions to strike the weakened invader. The objective is to cripple or destroy the attacker's main force, capture or eliminate leaders, and deny support units and supplies. The defender then eliminates the remaining opponent, who is extended, surrounded, and demoralized. During the counterattack, however, commanders must avoid overextending themselves.

So, while defensive strategies are stronger with passive objectives, the attack is weaker with active objectives. Success in defense occurs when the defender achieves their objectives, which may include deterring or repelling an attack, inflicting heavy losses, protecting key assets, buying time, or enabling favorable negotiations.

The seven keys to a powerful defense are:

1. Preparation: Maximize available time and resources to prepare fortifications, obstacles, equipment, and personnel.
2. Security: Deny enemy reconnaissance while securing early warning of opposing movements and intentions.
3. Mass effects: Synchronize and coordinate fires at decisive points; use overlapping fields of fire and mutual support to overwhelm attackers.
4. Flexibility: Maintain a mobile reserve and alternate positions to respond to enemy actions and retain adaptability.
5. Appropriateness: Tailor defenses to terrain, available resources, and the enemy's likely tactics.
6. Sustainability: Protect supply lines and stockpiles to ensure prolonged defensive capacity.

7. Deception: Employ camouflage, decoys, and misinformation to conceal true strengths, intentions, and positions.

Let's consider some defensive strategies.

DEFENSE-IN-PLACE

A defensive strategy may focus on fortifying *strongholds* or *fortifications* and awaiting an opponent's attack. The objective is to build a powerful defense that either deters an aggressor from attacking, blocks an aggressor from accessing key terrain, or defeats an aggressor.

There are two forms of a defense-in-place strategy. The first, an indefinite defense, involves holding an area without a defined end date, such as securing a national border. The second, a blocking mission, focuses on denying an opponent access to an area or route for a limited time or until a specified event occurs, such as delaying an enemy until friendly forces have repositioned or completed a withdrawal.

This strategy also offers a variety of deployment options. A linear defense array forces in a line, perpendicular to an enemy's expected route, massing military power against the greatest threat. While effective against direct assaults, such as along a contested border, it leaves fewer defenses in other sectors and is poorly suited against opponents that can bypass fixed positions or that employ area-effect weapons, such as nuclear, orbital, chemical, or biological weapons, that can destroy concentrated forces.

A perimeter defense deploys forces in all directions to form a defensive ring. This strategy aims to protect against attacks from any direction but is weaker against massed attacks from a single direction. It is most effective when the defender doesn't know which direction the enemy might come.

Regional strongholds create pockets of resistance, such as encampments, firebases, and orbital bodies. These strongholds may be mutually supporting, reinforcing each other, serving as outposts that provide early warning and fallback positions, or holding out for rapid response forces. Defenders may use remote outposts to stand guard, gather intelligence, or win the natives' hearts and minds. Regional strongholds disperse forces and reduce the effect of devastating area weapons.

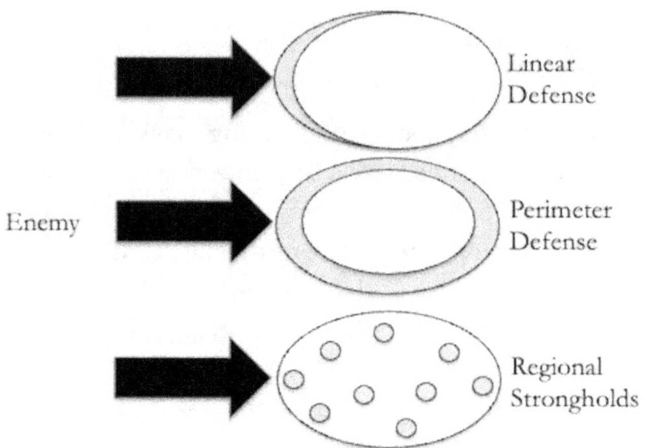

An attacker (black arrow) encountering different defensive strategies.

Fortifications provide many benefits to defenders beyond protecting the occupants. If a society establishes them on a border close to a neighbor, ostensibly as a defensive base, they can also use them to build-up forces covertly before a surprise invasion of a neighbor, to store supplies in case of a siege or an invasion, and as a visible sign of the society's military power and their political leaders' might (Vauban and Rothrock 1740).

The Death Star battle stations of **Star Wars** are mobile regional strongholds and highly visible symbols of Imperial might, developed to instill terror in the Emperor's opponents. Accordingly, the Rebels' destruction of these mobile fortresses becomes a visible sign of Imperial weakness and a recruiting tool for the Rebellion.

DENIAL

A denial strategy seeks to prevent an adversary from achieving military objectives or waging war by physically obstructing their efforts.

Unlike deterrence strategies that rely on threats of retaliation to coerce an opponent not to perform some action (discussed in Chapter Nineteen), denial focuses on direct actions to prevent an opponent from taking or using something.

Destroying bridges that span a wide, raging river may deny an invader access to the other side, severing their potential route of advance and disrupting their momentum, or channeling them onto a path of the defender's choosing.

There are many other forms of denial operations, including cyberattacks that disable information networks, electronic warfare that jams or deceives sensors, and territorial denial through mines, orbital defenses, or radiation zones. Other methods involve destroying or contaminating critical resources, infrastructure, or supply chains to halt production or movement. Intelligence denial, diplomatic isolation, legal constraints, and psychological operations can prevent an adversary from acting or diminish their will to fight. Even environmental manipulation, such as sabotaging terraforming efforts, can render territory unusable. All denial strategies exploit dependencies to prevent an opponent from achieving their objectives without requiring direct confrontation.

Scorched Earth tactics exploit Maxim 42, deny an enemy what you cannot possess, by burning crops, slaughtering livestock, destroying structures, contaminating terrain or supplies, and removing all usable resources to leave nothing for an opponent. This approach is particularly effective against an invader who relies on scavenging to sustain their offensive, because deliberately denying resources weakens enemy strength, slows their advance, undermines morale, and may compel them to abandon the campaign.

A denial strategy is rarely successful without significant effort and commitment. Denied areas become obstacles, and like all obstacles, an adversary will eventually overcome them unless they are kept under constant observation and fire. Most often, denial is employed to weaken an opponent during the initial phase of an exhaustion strategy, followed by a counterattack to destroy the now-impotent adversary.

That said, denial operations carry risks. If a defending force must retreat through or re-enter the area it devastated, it may suffer the same deprivation it intended to inflict on the enemy. In such cases, the defender becomes a victim of their own denial strategy.

MOBILE DEFENSE

Rather than placing forces into fixed positions that an adversary could target, overrun, or bypass, a defender may use a mobile defense strategy to trade space for time. Its objective is to slow, weaken, and frustrate the attacker until it halts or reaches its culmination point, at which point the defender can defeat the offensive. Using this strategy, a defender falls back as needed, choosing the optimal time and place for defensive battles and counterattacks on terrain that provides advantages. The defender uses an attrition strategy to wound and delay the attacker while avoiding becoming decisively engaged or encircled. This approach conserves military power, inflicts casualties, and complicates an attacker's plans, while reducing the defender's vulnerability to concentrated strikes or weapons of mass destruction.

A form of mobile defense is the elastic defense, also known as defense in depth, in which a defender uses a series of defensive zones to absorb an attacker, then counterattacks to repel and destroy the weakened invader. While forward forces delay the enemy, the defender can construct defenses, fortifications, and routes within the defensive area. Once the delaying action is complete, units fall back to prepared positions for the next battle, while additional fallback positions are established further to the rear. All the while, the defender prevents being surrounded, decisively engaged, or bypassed.

Once again, applying Maxim 42, the defender strips a surrendered region of anything useful, including weapons, supplies, or labor. The defender may also conceal supply caches for use during a counterattack.

As a defender withdraws, it may consider purging the countryside of anything of value. If it is unable or unwilling to evacuate civilians from the combat zone, it will be leaving them at the mercy of invading forces. Such purges can harm the population, who may feel betrayed and abandoned without food or supplies. Civilians often interpret this tactic as prioritizing military objectives over civilian lives, eroding public trust, and producing long-term political consequences. Commanders should consider evacuations, protected corridors, and targeted denial measures that spare noncombatants.

A military force may also leave spies, scouts, and guerrilla forces in abandoned terrain to provide intelligence and perform unconventional attacks to frustrate and wear down the invaders. In this way, the defender can fall back, reconstitute, rebuild, and rest, while the attacker is constantly harassed by the forces left behind.

RETROGRADE

"Run, Luke, run!"

— OBI-WAN KENOBI, JEDI MASTER (LUCAS 1977)

Retrograde operations are deliberate, organized movements away from an opponent, intended to preserve military power or reposition forces. There are four types of retrograde operations: withdrawal, retirement, breakout, and delay. A withdrawal involves breaking contact with the enemy and moving away while still engaged. Forces may withdraw to consolidate their strength, occupy more defensible ground, bait and ambush a pursuing enemy, or evacuate a region.

A retirement is a maneuver away from an enemy when not in contact. It is often used to regroup or resupply, conserve military forces threatened by a stronger opponent, or lure an opponent into a trap.

A breakout is a maneuver to avoid or escape a flanking or encircling enemy. Sometimes this is merely outmaneuvering an opponent. Other times, this may involve maintaining the illusion of being stationary, while one group creates a hole through the enemy that the rest of the force uses to escape.

A delay is a retrograde operation intended to slow a pursuer's advance through deliberate and controlled actions. Unlike a withdrawal, a delay is a calculated effort to disrupt the enemy's momentum and timing, forcing them to pause for resupply and reorganization and buying critical time for friendly forces or civilians to escape, regroup, or prepare for a counteroffensive.

Ultimately, the purpose of a delaying action is not to achieve victory, but to hinder the enemy's advance and, in doing so, prevent the adversary from winning.

Maxim 49: Decline a fight until the odds are in your favor. Retrograde when the benefit of moving outweighs the cost of staying.

A successful retrograde can create tremendous opportunities, but failure risks catastrophic consequences. Success depends on maintaining control and discipline throughout the operation. Once friendly forces begin moving away from the enemy, cohesion may falter, and a deliberate, controlled maneuver can quickly devolve into a chaotic *rout*. As discipline collapses, units break into individuals focused solely on survival. In this state of panic and desperation, the enemy may relentlessly pursue and destroy scattered, exhausted, and demoralized remnants. Within the chaos of the rout, armies expose themselves to the very real threat of annihilation.

Commanders can employ a variety of tactics to maximize the effectiveness of a retrograde. They may use deception to create the illusion that forces remain in place, to obscure the actual route, or to suggest greater strength than exists. Security measures must prevent surprise attacks, including guarding against being flanked and bypassed. Commanders can emplace obstacles, such as minefields and physical barriers, and deploy obscurants, like smoke generators or fires, to hinder pursuers and disrupt enemy tracking. Forces may deploy snipers, ambushers, harassing forces, mines, and booby-traps to slow and demoralize pursuers. Finally, defenders may deny pursuers resources by destroying or contaminating supplies, food, and water.

Corollary of Maxim 13: While harassment can be a potent tool, victories are achieved through offensive actions.

A commander may deploy a *rear guard* that maintains contact with the enemy while the main body withdraws. Its mission is to trade space and effort for time by delaying the pursuer, disrupting their momentum, and inflicting casualties, without becoming *decisively engaged*.

Although often associated with tactical maneuver, a retrograde can also serve as a deliberate strategic operation. At the strategic level, retrograde actions trade territory for time, preserve combat power, draw an adversary into overextension, or reposition forces to more favorable political, geographic, or logistical conditions. Such decisions are not signs of defeat but calculated measures that safeguard a society's resources, maintain legitimacy, and set conditions for future offensive action. Examples include evacuating a capital or principal city to preserve population and government continuity, evacuating trained forces and leadership, and withdrawing to interior lines to shorten logistics and strain enemy sustainment. When executed intentionally, a strategic retrograde can deny an adversary decisive victory and shape the outcome of a war as decisively as a successful advance.

When a commander detects an enemy in retrograde, they may be tempted to exploit the opportunity by transforming the withdrawal into a rout. This demands audacity and aggression, pursuing the opponent, maintaining contact as they flee, and often committing reserves to sustain pressure. The objective is not only to pursue, but to collapse the retreat entirely.

Pursuers can apply relentless pressure and strike at weak points such as rear guards, flanks, or supply lines. Speed and maneuver are critical; cutting off escape routes or seizing chokepoints may force a decisive engagement or surrender. Harassment from long-range fires keeps the retreating force off balance, while mobile units bypass defenses to strike deeper elements. A well-executed pursuit can seize the initiative, reclaim the battlefield, inflict heavy losses, demoralize, and ultimately destroy the withdrawing force.

However, the pursuit is fraught with risk. A force focused on exploiting the enemy's weakness may advance too far, outpacing its supplies, its support, and the protective coverage of friendly forces on its flanks and in reserve. At any moment, the withdrawing enemy may halt and "go to ground," regroup, and turn back to present a unified front that catches disorganized pursuers off guard. A retrograde may even be deliberate, serving as a feint to lure pursuers into an ambush that turns a bold pursuit into disaster.

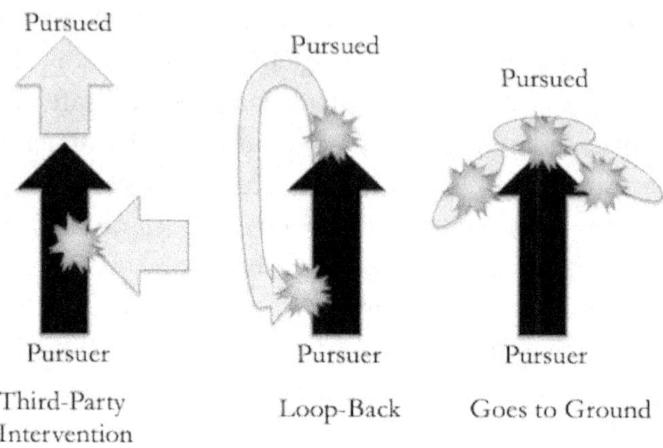

Tactics to ambush a pursuer (in black).

As a pursuit pushes deeper into enemy territory, the pursuer's vulnerabilities compound. Flanks become exposed, supply lines stretch thin, and sustaining food, ammunition, and rest grows more difficult with distance. Continuous operations wear down troops and equipment, leading to fatigue, low morale, and decreased combat effectiveness. Soldiers pushed to their limits may become less responsive to orders and more prone to mistakes.

A reckless commander, blinded by momentum, may advance too far in the pursuit and risk collapse or a surprise counterattack. Commanders must therefore temper aggression with caution. Success depends on accurate intelligence, knowledge of the terrain, and firm control of formations. A disciplined pursuit can break the enemy; an undisciplined pursuit can break the pursuer.

Maxim 50: Few things are more dangerous than a pursuit.[1]
Pursuing an opponent is a calculated risk that can lead to victory or defeat.

In *Star Wars,* the Rebels conduct a delaying action during the Battle of Hoth. Confronted by the Empire's overwhelming firepower and numbers, the Rebels employ a combination of defenses to slow the enemy advance. A planetary shield generator prevents orbital bombardment, an ion cannon disables Imperial ships to clear escape routes, and ground troops hold the line against a ground assault. These coordinated efforts are not intended to win the battle, but to buy precious time for Rebel personnel to escape the planet before the Imperial forces overrun the base.

While the Rebels ultimately lose their base, they preserve most of their troops and vessels. Their sacrifice allows Luke Skywalker to slip through Darth Vader's grasp and head to Dagobah to begin his Jedi training, ultimately leading to the downfall of Emperor Palpatine (Lucas 1980).

A *forlorn hope*, also known as a last stand, is a defensive operation in which the defenders face overwhelming odds and are likely to suffer significant or total casualties. While a forlorn hope often ends in the defender's capture, injury, or death, their sacrifice often delays a pursuing or invading force and enables others to escape.

Corollary of Maxim 25: Desperate circumstances may require desperate measures.

A forlorn hope, despite its grim outcome, can have a profound psychological impact on both sides. The heavy casualties it may inflict upon the attackers reveal their vulnerability, perhaps shaking their sense of invincibility. The defenders' sacrifice may become a rallying point, a symbol of courage and resistance that may inspire others and harden the collective resolve. Even in defeat, such an act can echo as a battle cry, reinforcing others' will to endure and fight on.

Obi-Wan Kenobi, the legendary Jedi Master of *Star Wars*, sacrifices himself in a last stand during a raid on the Death Star battle station. After sabotaging the Death Star's tractor beam and freeing a political prisoner, the Rebel infiltrators are forced into a running gunfight with Imperial forces. Kenobi becomes decisively engaged in a hangar bay when he encounters Darth Vader, his former Jedi apprentice turned

Imperial sociopath. Realizing he can't escape, Kenobi engages Vader in a delaying action that distracts nearby Imperial Stormtroopers. Vader cuts him down, but his sacrifice enables the remaining Rebels to escape from the Death Star.

FEINT

> "You're outmanned, you're outgunned, you're out equipped. What else have you got?" Commander Riker asks Worf as they face a seemingly hopeless mission.
>
> Worf smiles. "Guile."
>
> — DISCUSSION BETWEEN COMMANDER WILLIAM RIKER AND LIEUTENANT WORF, USS ENTERPRISE (NCC-1701-D) (KEMPER 1989)

A feint is a deliberate deception intended to divert an opponent's attention from the true objective. Its purpose is to lure the enemy into a disadvantageous position and then strike. The bait may take many forms, such as a display of weakness, feigned injury, apparent disinterest, or the promise of valuable targets, such as treasures, key leaders, or vital resources. A defender might use a small force to mimic the main body, enticing its opponent to go elsewhere, only to have the actual main body strike unexpectedly, where the opponent is not ready. Likewise, an attacker may feign a withdrawal to lure a defender out of its fortifications and into a kill zone, where it is crushed in an envelopment. Or a defender might entice their opponent to advance beyond its culmination point before springing a trap.

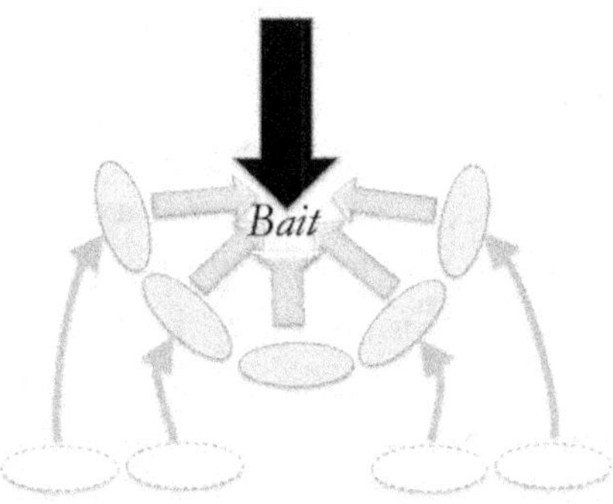

An enemy (in black) pursues bait and is surrounded by defending forces
(in gray).

A feint is especially effective against an arrogant opponent who believes themselves invincible. When the deception is revealed and the trap closes, survivors may suffer more than material loss. The shock of betrayal and the humiliation of being outwitted can shatter confidence in their leaders and cause them to question their cause, undermining morale and cohesion.

Guerrillas use a version of the feint when they hit a target and then withdraw, drawing pursuers into ambushes or sniper attacks designed to demoralize and inflict losses. Alternatively, a small force can use a feint to entice a security, reserve, or response force away from its post, leaving key resources unguarded. While the pursuers are delayed by traps, snipers, or ambushes, a raiding party can seize or destroy the unguarded resources, compounding the enemy's confusion and losses. This may be especially effective in terrain that limits visibility or movement, such as forests, cities, or mountains, where the enemy's pursuit can be easily funneled or delayed.

ACTIVE DEFENSE

A society may view its neighbor's alliances or military buildup as a threat and opt to launch a preemptive strike. This strike may aim to neutralize the threat, rather than allowing its neighbor to continue its preparations and choose the time and location to strike. A preemptive strike to disrupt, delay, deter, or destroy an opponent's military capabilities with no intention of occupying its territory is called an *active defense* or spoiling attack.

Striking an opponent at an unexpected time and place embodies Maxim 21: Strike when an opponent is vulnerable. By striking first, a society may minimize the risk to its own forces while simultaneously disrupting a neighbor's preparations and perhaps preventing an imminent attack. This aggressive and unforeseen strike may generate psychological and political shock, destabilize the opponent's military, leadership, and populace, and deter other parties from joining or supporting the adversary's cause

Spoiling attacks are not without risk. In the *Star Trek* universe, the aliens Cardassians mistakenly believe that a Federation outpost on the planet Setlik III is a staging ground for a massive Federation attack. The Cardassians preemptively attack and massacre the nearly one hundred civilians on the outpost (Okuda, Okuda, and Drexler 1999), horrifying the Federation, and starting a bloody, two-decade-long galactic war that snuffs out millions of lives (Moore 1994). Here, an active defense is a military success that results in a political and moral failure.

Some may view an active defense as indistinguishable from a preemptive strike that breaks the peace on the assumption that a neighbor is about to attack. Such actions blur the line between defense and aggression, especially when based on uncertain intelligence or subjective interpretations of an adversary's intentions. A state claiming self-defense may, in practice, initiate hostilities and thus appear as the aggressor in others' eyes. This perception can erode diplomatic credibility, invite condemnation or retaliation, and escalate a conflict that might otherwise have been avoided.

CONCLUSION

The art of defense is not merely about withstanding an opponent's attack but about shaping the battlespace to deny the adversary the initiative, preserve freedom of maneuver, and create opportunities for exploitation.

Astute leaders recognize that rigid defensive strategies are rarely ideal. Effective defense, like offense, demands adaptability: the ability to adjust, adopt, or integrate various tactics in response to evolving circumstances. An astute leader may employ a feint to provoke a neighbor into launching a preemptive spoiling attack. In doing so, the attacker abandons their fortified positions and exposes themselves to a well-planned counter-offensive involving harassment, ambushes, and decisive engagements. When executed with precision, this may simultaneously defeat their opponent and cast the defender as the victim of unprovoked aggression, garnering sympathy and support from other neighbors and allies.

The next chapters delve deeper into the complex dance between diplomacy and force, exploring the intricacies of intimidation, subversion, and various political strategies. Together, these strategies shape the balance of power and influence across contested domains.

CHAPTER 19
INTIMIDATION

"Fear will keep the local systems in line. Fear of this battle station."

— GRAND MOFF WILHUFF TARKIN, GOVERNOR OF
THE IMPERIAL OUTLAND REGIONS (LUCAS 1977)

Peaceful conflict resolution, aimed at achieving objectives without violence, can reflect superior strategy and leadership. It conserves resources and manpower, which is crucial in prolonged conflicts or when resources are limited. It also helps maintain public support and sustain morale among both military and civilian populations, factors vital for long-term success. Diplomacy addresses the root causes of disputes, fosters common ground, builds relationships, and strengthens long-term stability. Moreover, using diplomacy and negotiation rather than armed conflict cultivates a more favorable image with neighbors. Most importantly, peaceful conflict resolution minimizes civilian casualties, infrastructure damage, and suffering.

Corollary of Maxim 10: True victory achieves objectives with minimal losses.[1]
Victory without bloodshed is more compassionate and sustainable, and results in less animosity.

Politicians may pursue their objectives through nonviolent means, such as persuasion and *enticement*. Alternatively, leaders may employ intimidation by threatening something their opponents value, while simultaneously minimizing risk to their own personnel and resources (Tzu and Cleary 1988). Intimidation can take many forms, including threats of public embarrassment or blackmail; however, this chapter focuses specifically on *coercion*, what some call the diplomacy of violence (Schelling 1966).

Maxim 51: Negotiate from a position of strength.
A warrior negotiates with a knee on their enemy's chest and a knife at their throat.

COERCION

"Perhaps she would respond with an alternative form of persuasion."

— GRAND MOFF WILHUFF TARKIN, GOVERNOR OF
THE IMPERIAL OUTLAND REGIONS (LUCAS 1977)

Threats of violence can serve as powerful tools of coercion, enabling leaders to compel others to comply with their objectives. These threats may target military, economic, political, cultural, or religious interests. The objective may be to fracture an alliance in a divide-and-conquer strategy or to force a neighbor to act in a particular manner.

Coercion depends on both parties making rational choices. For coercion to succeed: (1) the target must be rational; (2) the proposed alternative must appear to serve the target's interests; (3) the threatened punishment must be credible and outweigh the cost of compliance; and (4) both parties must understand each other's reasoning and values. If any of these conditions fail, coercion may prove ineffective or even have the opposite effect.

Two coercive strategies are *deterrence* and *compellence*. Deterrence aims to prevent an adversary from taking a specific action by threatening punishment if the action occurs. The strategy relies on waiting to see whether the threat alone is enough to influence behavior. Deterrence succeeds when the threat of punishment persuades the opponent not to perform the trigger action. However, if deterrence fails, punishment may become necessary to preserve credibility and enforce compliance.

Successful coercion requires continuous monitoring of the target's behavior. For example, a threat to halt trade shipments must be backed by the capability to observe and verify compliance, including activities within illicit markets. Without effective oversight, coercion becomes vulnerable to ambiguity, misunderstandings, or deliberate manipulation by the target.

The aim is not to harm the opponent, but to dissuade them from performing specific actions. Success depends on the opponent's belief that the threatener possesses both the capability and the will to carry out the threatened punishment. Public demonstrations of weapon capabilities and a consistent history of following through on threats can reinforce this belief.

Importantly, the threatener need not perform the punishment or even possess the means or intention to do so. Success only requires that the target believes the threatener can and will act. A rational opponent, in the human sense of the word, may avoid an action if they believe the cost of punishment outweighs the benefit. Yet, politics and war are rarely governed by pure logic. Emotions, pride, and ideology can undermine rational calculations, making deterrence a less reliable strategy.

Deterrence is a passive strategy. It seeks to preserve the status quo, and in doing so, allows both sides time to build strength, seek alliances, and harden or evacuate potential targets. But if the opponent disregards the deterrent and performs the trigger action, they are challenging the instigator to respond. In this way, a leader using deterrence often cedes the initiative to their opponent.

Unlike deterrence, which seeks to prevent action, compellence is an active strategy intended to force an opponent to alter their behavior. It depends on the application of sustained punishment, imposed until the cost of continuing exceeds the perceived benefit, and the target changes their behavior. The objective is not to preserve the status quo, but to impose compliance with an alternative course of action.

Compellence involves a shared responsibility between the coercing and the coerced. The initiator bears responsibility for sustaining pressure, often at considerable cost, until the adversary alters their behavior. Either side can end the exchange: the initiator by ceasing the coercive measures, or the opponent by complying with the demands.

Raids are often effective tools of compellence. By gradually escalating the scale or intensity of limited actions, a force can pressure an adversary to change its behavior without incurring the costs of full-scale open warfare. Small guerrilla groups and criminal organizations employ targeted violence to coerce stronger powers, leveraging their agility, lower risk exposure, and a propensity for bloody, attention-grabbing acts to shock adversaries into compliance.

The challenge of coercion lies in understanding how an opponent weighs potential costs and benefits. Coercive strategies often target a society's forms of power, such as its population, economy, leadership, or military strength, in order to erode confidence in the government's ability to provide protection or to weaken a leader's control over the population.

While the direct costs and benefits of coercion, such as troop deployments or economic sanctions, are often easier to quantify, the indirect aspects can be far more elusive. Misjudging these can produce unintended consequences. For instance, threatening to destroy a revered religious site or a society's capital could provoke a spectacular backlash. Instead of the expected submission, it might galvanize the target population into a fierce, unified resistance.

Therefore, intelligence is vital to understanding a target's cultural values and potential responses. However, acquiring accurate, actionable insight is significantly more difficult when dealing with foreign cultures or species, where traditional methods of intelligence gathering and analysis might prove inadequate. In such situations, accurately calculating the true costs and benefits becomes a formidable task, prone to error.

Coercion is effective only when a society comprehends the threat, the conditions attached to it, and the consequences of noncompliance. Maxim 16 cautions that coercion fails when societies cannot understand one another, or when it is directed at groups that reject negotiations altogether, such as zealots and fanatics. Similarly, conflicts centered on the survival or eradication of a philosophy or species rarely permit compromise.

Even when coercion compels action, it may only yield short-term compliance. Maxim 19 observes that coercion produces obedience rather than loyalty. A resulting uneasy peace may lead to an arms race and heightened tensions. Over time, the balance of power may change, prompting the coerced party to strike preemptively to neutralize the threat. In the end, the slave may seek to become the master.

Corollary of Maxim 19 and 51: Obedience at the end of a bayonet breeds dissent.

Societies respond to threats in varied ways, shaped by their culture, political institutions, and the perceived severity of the threat. A warrior society may favor swift retaliation or preemptive action to eliminate the source of intimidation, emphasizing strength and decisiveness. By contrast, a diplomatic culture may seek to defuse a threat through negotiation, alliance building, or indirect influence. Some societies may appear outwardly passive, ignoring the threat while quietly preparing a calculated response, such as establishing traps or ambushes to punish aggression on their own terms. Others may work to reduce the threat's power by forming coalitions, isolating it politically, or undermining its credibility or legitimacy. Still others may target the threat's critical support systems and dependencies, including key individuals, infrastructure, or logistics, to weaken its capacity to act. These responses reflect more than just tactical judgment; they express a society's values, concepts of honor, and understanding of what it means to confront danger.

In the *Star Wars* universe, the Emperor believes fear is the most effective means of governing the galaxy and maintaining peace. He entrusts Grand Moff Wilhuff Tarkin with the Death Star, a massive planet-killing weapon intended to project imperial power and coerce obedience on a galactic scale (Lucas 1977). Tarkin declares to the captured Rebel leader Leia Organa, that the Death Star is the ultimate weapon, ensuring that "no star system will dare oppose the Emperor." Leia cautions him that, "The more you tighten your grip, the more star systems will slip through your fingers" (Lucas 1977).

Before this confrontation, the Rebel Alliance initiated covert action to neutralize the Death Star program at its source. Rebel forces move to extract or eliminate the scientists responsible for developing the Death Star and its powerful weaponry. Although these efforts failed to stop the program outright, Jyn Erso leads an operation to recover the Death Star plans, successfully revealing a critical vulnerability in the weapon's design (Edwards 2016). This discovery not only provides a means to counter the superweapon but will inspire the Rebellion by demonstrating that the Empire's most powerful instrument is not invulnerable.

Unaware of this compromise, Tarkin proceeds with a demonstration of terror, threatening to destroy Organa's home planet of Alderaan unless she reveals the location of a Rebel stronghold. She names the planet Dantooine, but Tarkin, unconvinced and intent on making an effective demonstration, orders Alderaan's destruction. He believes this display of the Death Star's destructive power, coupled with the emperor's resolve, will coerce the galaxy into submission (Lucas 1977).

Instead, the destruction of Alderaan strengthens the Rebels' resolve and inspires others to join their cause. Although the massively powerful Death Star represents unmatched strategic power, the Empire's use of it backfires, producing severe and unintended strategic consequences. Rather than instilling fear, the Empire's display of terror highlights its brutality, transforming outrage into unity among its enemies.

CONCLUSION

This chapter has examined intimidation strategies, in which the projection of power, manipulation of perception, and deliberate cultivation of fear serve as primary instruments of influence. Such a strategy can shape relationships, advance political objectives, and, sometimes, prevent armed conflict. Yet, as illustrated in *Star Wars*, intimidation carries significant risks. Rather than eroding an adversary's will to resist, it may harden opposition, strengthen resolve, and unite otherwise fragmented groups against a common enemy.

The following chapter examines subversive warfare. Whereas intimidation depends on overt demonstrations of power, subversion operates in the shadows.

CHAPTER 20
SUBVERSION

"You can't stop the signal."

— MR. UNIVERSE, SUBVERSIVE (WHEDON 2005)

A faction may lack the capability to confront its adversary through direct military action. This situation arises when a subgroup within a society is too weak to bring about meaningful political change, or when an occupied population lacks the strength to expel a foreign power. In either case, the disaffected and disenfranchised may come to view violence as the only viable way of resisting what they perceive as tyranny and oppression. Alternatively, a society may attempt to compel a neighbor to change without resorting to open warfare by employing a *subversive* strategy that gradually exhausts, undermines, and ultimately overthrows the opponent.

Conventional conflicts usually involve belligerents with comparable means and ways, resulting in symmetric wars in which both sides employ similar strengths and methods. When confronted by a weaker opponent, an established order often seeks to maintain control and resolve the conflict quickly, either through concessions or by forcing a decisive battle. In such circumstances, its professional military, superior in training, discipline, organization, weaponry, and tactics, can readily overwhelm the opposition.

However, subversive groups often lack both the means and the desire to directly confront an established order. Maxim 21 infers concentrating forces against weaker opponents to achieve swift victory, and that a poorly trained, lightly equipped force stands little chance in a direct confrontation with a professional army. As a result, subversives avoid decisive engagements and instead seek to weaken or exhaust the dominant power. They rely on different means, employ different ways, and pursue different objectives, giving rise to *asymmetric wars*.

Unlike traditional offensive strategies of annihilation or exhaustion, subversion is a defensive exhaustion strategy. The subversive force begins as the weaker party, unable to control the battlespace, but as it gains power and weakens its adversary, it can eventually deny the opponent unrestricted access to the terrain. Maxim 49 cautions against engaging in battle until the odds are favorable. The subversive's advantage lies in exploiting moments of *local superiority*, striking vulnerable targets with surprise and precision. Still, they must remain mindful of their strategic inferiority. Though the temptation to retaliate or demonstrate courage may be strong, guerrillas who engage too soon, before their enemy is weakened or overextended, risk annihilation. When the enemy advances in strength, guerrillas must withdraw, disperse, and disappear.

In *Dune*, the Harkonnens, unable to openly challenge House Atreides after the Emperor's decree transferring control of Arrakis and its immense wealth, resort to a layered campaign of deception, sabotage, and infiltration. When the Harkonnens withdrew from the planet, they left behind *booby-trapped* equipment to cripple Atreides operations and cause immediate setbacks. They plant an assassin within the palace to target the Atreides heir, aiming to destabilize leadership and sow fear. Most significantly, they subvert the Atreides from within by coercing Dr. Yueh, the trusted Atreides family physician. His betrayal disables the palace defenses and incapacitates Duke Leto at the critical moment.

This operation exemplifies a subversive strategy. Rather than confronting a strong opponent through open warfare, the Harkonnens rely on betrayal, psychological disruption, and sabotage to erode their enemy from within, softening the target before a swift and devastating coup.

The key to subversion is understanding the populace, politicians, the economy, and the military situation. The subversive then strikes an opponent's weakness, which is often a critical component. For example, relatively weak subversives may threaten a democracy by targeting the voting populace to undermine a politician's hold, disrupt the political process, and persuade their foe to sue for peace. Likewise, violent surprise attacks against military members fighting a "pointless war" may convince citizens to end the violence at any cost.

Corollary of Maxim 21: Subversives undermine resolve, not means. Subversives conserve their strength by adopting ways that reduce an adversary's commitment to a conflict.

Powerful conventional forces often view subversive strategies as immoral or illegal, accusing subversives of not "fighting fair." But "fairness" is a subjective notion, usually defined by those in power. As Leia Organa notes, "In war, you can do just about anything the other side can't stop you from doing" (Zahn 1993).

Political and military leaders often describe strategies that challenge the status quo and rely on "unconventional" or "irregular" methods using vivid terms such as operations other than war, armed propaganda, insurrections, and revolutions. For clarity, however, these operations can be grouped into four subversive strategies: *undermining, overthrow, betrayal,* and *decapitation.*

UNDERMINING

"If you can create sufficient fear in your enemies, you may not have to fight them. Always remember that terror is also a form of communication."

— DELENN, MINBARI AMBASSADOR TO BABYLON 5
AND A FOUNDER OF THE INTERSTELLAR ALLIANCE
(STRACZYNSKI 1998B)

Undermining involves deliberately weakening an opponent, often through indirect or deceptive ways. This strategy offers a cost-effective, low-risk approach to expansion by eroding the targeted society from within. It typically begins with *covert* operations aimed at sowing discord, fostering distrust, and inciting unrest among political factions, social classes, ethnic groups, or other societal segments.

Subversive efforts may focus on destabilizing the society itself through the spread of harmful influences, such as recreational drugs; undermining morale with terrorist attacks against civilians; or attacking cultural cohesion by inflaming internal divisions. Psychological operations often reinforce physical actions, including social media manipulation, provocative news coverage, polarizing political rhetoric, and the orchestration of civil protests.

In the universe of *Dune*, the Bene Gesserit sisterhood employs subversion to pursue its long-term objectives and ends. Through manipulation, espionage, and selective breeding, the sisterhood influences the political and social structures of many worlds, particularly Arrakis. Their ultimate end is the creation of the Kwisatz Haderach, a super-being endowed with immense power and knowledge. To achieve this, the Bene Gesserit infiltrate and discreetly steer the course of various political and religious movements across the galaxy, shaping events and leadership to serve its overarching design.

To achieve their objectives, societies may expand military operations beyond direct confrontation with an adversary's armed forces. Instead, they may strike at other forms of power by undermining elections, manipulating markets, sabotaging critical infrastructures, stealing military and industrial secrets, and reducing a population's quality of life. Such actions can erode confidence in leadership, weaken national unity, and lead the populace to believe their government is incapable or unwilling to protect them.

The ultimate objective of subversive efforts is to mobilize popular support and political influence against their opponent. By expanding their influence and means, subversives aim to trigger a cascading collapse of their adversary, clearing the way for greater control and dominance.

Subverting a neighbor may appear to be a low-risk endeavor until the victim asserts its right to determine its own destiny. The victim will regard this meddling as a threat to their sovereignty and may punish interlopers to deter others from interfering in its internal affairs. Escalation may take the form of economic sanctions, covert retaliation, proxy wars, or open military responses, each carrying broader consequences.

OVERTHROW

"The Jedi have tried to overthrow the Republic."

— ANAKIN SKYWALKER, JEDI KNIGHT (LUCAS 2005)

An *overthrow* is the use of force by a small group within or outside a society to depose an established government. While traditional warfare involves one society attacking another, an overthrow may stem from internal dissent, subversion, or a combination of both.

Overthrows come in various flavors, each based on its distinct end. A targeted overthrow might aim to depose a single leader, perhaps with a predetermined successor in mind. This often seeks to maintain the existing system of government while simply replacing the person at the helm. More ambitious is a regime change, in which the subversive seeks to remove not just a single leader but to dismantle the existing power structure and replace a significant portion of the ruling elite. The most radical form of overthrow seeks systemic transformation. It changes not only the leadership or the ruling party; it fundamentally alters the political and social order, reshaping the very fabric of society, rather than replacing who holds power.

Overthrows often involve relatively weak subversive *guerrillas*[1] fighting powerful militaries in protracted harassments. They do not seek immediate battlefield victories, but to exhaust the opponent until political change becomes inevitable.

Corollary of Maxims 13 and 18: Intelligence is the foundation of guerrilla operations and survival.

Guerrillas often seek legitimacy through propaganda and political messaging, framing their cause as just and necessary, to recruit followers and build a sympathetic base. At the same time, they employ violence to eliminate competition, coerce their opponents, destabilize institutions, and exploit divisions within society. By doing so, they aim to fracture the social and political order, making it more vulnerable to internal collapse.

Guerrillas thrive on asymmetry: avoiding direct confrontation with superior forces and relying on ambushes, sabotage, assassinations, and raids to inflict steady, demoralizing losses. They bleed their stronger adversaries over time, draining resources and public support.

Explosives and booby traps are among the guerrillas' most favored weapons. These devices instill fear, erode morale, and through their visible and unpredictable nature, create a constant sense of insecurity among occupying or defending forces. Since the guerrillas need not be present when such weapons are triggered, they avoid direct confrontation with their stronger opponent. Their anonymity makes them difficult to trace, while the low cost and ease of deployment minimize risk to the attacker.

Though often outnumbered, guerrillas maintain an advantage by avoiding decisive battles. They apply intelligence, mobility, and surprise to achieve local superiority and gradually wear down a larger opponent. Through a combination of ideology, violence, and calculated disruption, they seek to undermine the legitimacy of established powers and position themselves as the authentic voice of resistance. They concentrate their forces only when and where they hold the advantage, then disperse. They must always remember that their survival depends on evading the trap of conventional warfare.

Guerrillas understand the wisdom of Maxim 7: any dependency is a vulnerability. They target the lifeblood of a conventional military by focusing on the opponent's logistics network. By striking "softer" targets within the supply chain, such as transportation hubs, fuel depots, or communication lines, guerrillas may cripple forces dependent on those critical supplies. A military that cannot receive food, fuel, ammunition, or medical supplies will rapidly decline in both capability and morale.

In *Andor*, insurgents avoid direct attacks on Imperial garrisons and instead conduct a train robbery targeting an Imperial payroll moving along a predictable route. The operation exploits the Empire's reliance on centralized logistics, fixed schedules, and bureaucratic routine. By penetrating the supply system, the guerrillas seize resources critical to Imperial control while exposing the fragility of an apparatus optimized for efficiency rather than resilience. Although the material loss is notable, the psychological and organizational effects are greater, forcing the Empire to divert forces, tighten security, and impose harsher controls that strain both its logistics network and its legitimacy. The Aldhani operation demonstrates how attacks on logistical lifelines can produce asymmetric effects, showing that undermining supply and confidence may be more decisive than defeating combat forces outright (Gilroy 2022).

However, guerrillas may take this concept one step further by treating anyone who supports the political system or its warfighters as a legitimate military target. A government's support network extends beyond the battlefield to include individuals, organizations, and infrastructures that enable its operations, including civilian suppliers, logistics personnel, contractors, and others who contribute to the functioning of military and governmental systems.

By striking at these essential dependencies, guerrillas achieve asymmetric effects with limited means. They avoid direct confrontation with enemy troops while steadily eroding their enemy's operational strength and capacity to sustain the conflict.

These attacks also serve as a coercive tactic. By demonstrating the vulnerability of essential resources and services, guerrillas apply pressure on political leaders to reconsider the war effort, increasing the likelihood of a negotiated settlement.

Corollary of Maxim 6: Anyone viewed as supporting a side may be a target.
In guerrilla warfare, the enemy's rear is the guerrillas' front.

By threatening the dependencies that keep warfighters and society running, guerrillas compel their opponent to provide security for its combat forces and for the systems that manage, collect, process, store, and transport those dependencies. Military forces may also be needed to protect population centers, politicians, government facilities, and other highly visible people and organizations, ensuring society can operate and maintain popular and political support.

As fear increases, the guerrillas may compel their opponent to go on the defensive and create heavily fortified urban sites, infrastructures, and military bases connected by supply lines. This transition to the defensive cedes the initiative to the subversive. The guerrilla then shifts to lightning attacks against small, isolated defensive posts and the lightly defended supply lines that connect them.

At this point, conventional military forces may still outnumber guerrilla forces. However, those forces are now dispersed throughout the battlespace, with most troopers committed to security and defensive duties in military cantonments, base camps, and along supply routes protecting critical dependencies. Guerrillas can devote nearly all their forces to offensive operations. A savvy guerrilla leader then uses intelligence, surprise, maneuver, and mass to create *local superiority* against small portions of the much larger opponent in a perpetual divide-and-conquer campaign.

Corollary of Maxim 21: Bleed a larger force until it concedes, capitulates, or can be defeated outright.

In *The Hunger Games*, Katniss Everdeen's reluctant rise as the Mockingjay becomes a powerful symbol of defiance against the Capitol's brutal regime. Her survival in the arena, acts of compassion, and refusal to play by the Capitol's rules resonate across the districts, many of which had long suffered systemic exploitation and fear.

As the Mockingjay, Katniss galvanizes a fragmented and oppressed population, transforming her image into a rallying point that fuels a coordinated uprising against President Snow's tyranny.

In *Babylon 5*, John Sheridan emerges as the central figure in a rebellion that transcends human politics and unites multiple alien civilizations. After President Clark seizes control of Earth by arranging the assassination of his predecessor and imposing martial law, Sheridan refuses to recognize the regime's authority. Operating from Babylon 5, a neutral space station, he builds a coalition of Earth's military defectors, alien governments, and underground resistance groups. Using both military force and strategic messaging, Sheridan challenges Clark's authoritarian rule, eventually leading to the liberation of Earth and the restoration of democratic governance.

These examples demonstrate how oppression and propaganda can ignite rebellion. Individuals who serve as symbols of resistance can unify scattered, diverse factions toward a shared objective, even in the face of overwhelming power. The longer such individuals control the stage, the greater their influence on an opponent's alliances, recruiting efforts, political institutions, citizen morale, and external support.

In *Rogue One: A Star Wars Story*, Grand Moff Tarkin admonishes Director Krennic, saying, "You have made time an ally of the Rebellion" (Edwards 2016). Tarkin recognizes that every moment that the Empire fails to act gives the Rebels time to expand alliances, recruit supporters, fortify political institutions, rally citizens, and gain more combat power. Tarkin holds Director Krennic accountable for delays in completing the Death Star; the longer the station remains unfinished, the more opportunity the Rebellion has to organize, plan, and strike in ways that could threaten the Empire's dominance, build momentum, and shift public perception. For a conventional force such as the Empire, time becomes a liability rather than a neutral factor. Guerrillas thrive when larger powers respond slowly; the more time they are given, the greater their ability to destabilize authority and to threaten even the most powerful weapons.

Inverse of Maxims 11 and 21: Conventional forces cannot allow guerrillas to steadily erode their strength.

Simmering tensions, dissent, and dissatisfaction within a society can erupt into internal conflicts, and the names applied to those conflicts carry significant weight beyond mere semantics; labels influence public perception, confer or deny legitimacy, aid recruitment, and help secure support.[2]

A subversive group may overthrow a government through rebellion, revolution, or civil war. A rebellion is an attempt by a part of the population to resist a perceived oppressive government. It often arises when citizens believe peaceful reform is impossible and that rebellion is the only path to justice or autonomy. The people are fighting for something other than control of the state. They resist and defy authority to seek concessions, such as freedoms, rights, or basic needs.

A revolution involves a group seeking to overthrow the existing government and establish a new political order. Where a rebel seeks to force change, they are content with the political establishment. The revolutionary wants to destroy the political establishment and its laws.

A civil war involves two or more recognized factions fighting for control of the government and its structure.

These terms are subjective, falling along a spectrum, and influenced by the legal and cultural systems of those in power. For example, a small group may begin a rebellion to demand change. As it grows or in response to events, a portion of the populace may decide that the incumbent government must go, sparking a revolution. As the revolution spreads and the incumbent government collapses, creating a power vacuum across the state, a civil war may erupt as groups fight for control.

While passionate guerrillas can ignite rebellion, toppling a society requires a broader strategy. Large-scale change hinges on mobilizing the populace and on altering the military's loyalty, and this path is fraught with peril. The most formidable obstacle to any uprising is the forces tasked with maintaining order, since the military and paramilitary forces, including police, often possess the firepower, cohesion, and organization to crush an insurgency. Guerrillas may seek to defeat these by force, but indirect methods that disband, take control, or neutralize them often prove more effective. This may occur from the top-down, when senior military leaders support the subversives' aims, or bottom-up, when common troopers sympathize with the subversives. Guerrillas may also work to split these forces from the regime through persuasion, propaganda, or by creating conditions that encourage desertion or defection.

Beyond geography, internal conflicts are distinguished by their intensely personal character. In an interstate war, a society may demonize an external opponent to justify violence, to inspire its military and civilians, and to excuse the sacrifices often required to win wars. In an internal war, belligerents still demonize their opponents as supporters of the "evil established authority," or as "traitors to legitimate authority." Internal conflicts force neighbors and families to choose sides, and what begins as a political dispute often becomes personal, emotionally charged, and brutal, and all this hatred and violence is contained within the borders of a single state. As a result, internal conflicts frequently produce disproportionately greater loss of life and deeper economic strife.

Maxim 52: Internal conflicts often result in unbridled bitterness and unrestrained violence.

Internal conflicts also do not end cleanly, even when one side achieves a decisive victory. Having been vilified and persecuted for their beliefs and identities within their own homeland, the vanquished may view the victors as evil and cruel, and any honor the victors pay to their fallen as insulting and provocative. Unless the defeated are eliminated, reconciliation remains the only path to healing the wounds on both sides. Without it, resentment and unrest continue to simmer beneath the surface, creating the conditions for renewed conflict and violence.

BETRAYAL

"Assume everyone will betray you, and you will never be disappointed."

— TOBIAS BECKETT, CRIMINAL (KASDAN AND
KASDAN 2018)

Trust is the confidence placed in a person, object, organization, or system to act in accordance with expectations. It is not mere hope; it is a conclusion drawn from evidence, consistent patterns, sworn oaths, and demonstrated behavior. Trust establishes reliance on others to fulfill specific roles, responsibilities, expectations, and obligations. This reliance creates a dependency, linking the well-being or success of one entity to the reliability of another. However, Maxim 6 cautions that dependencies carry inherent risks, as failure or betrayal can cause significant harm or loss.

War reveals both the noblest and darkest aspects of people. In the face of chaos and danger, acts of courage, sacrifice, and loyalty often emerge as individuals risk their lives to protect others and to uphold principles. Yet war also magnifies fear and desperation, driving some to self-preservation or self-serving behavior at the expense of honor and morality. These pressures may cause individuals to betray a trust, whether by abandoning allies, revealing secrets, or choosing personal gain over the collective good.

**Corollary of Maxim 6: Trust enables betrayal.
You can only be betrayed by someone you trust.**

In the *Star Wars* universe, the Emperor's henchman, Darth Vader, arrives at Cloud City with a detachment of stormtroopers, and he coerces Lando Calrissian, the city's administrator, into betraying his friend Han Solo. In exchange, Vader promises to overlook Lando's criminal activities and to withdraw Imperial forces from the city when he departs. When the trap is sprung, Han and his compatriots regard Lando's actions as treachery. Lando, however, believes he had no choice given Vader's homicidal methods for handling dissent and the Empire's recent annihilation of the planet Alderaan and its nearly two billion inhabitants (Bray, Horton, and Barr 2017).

Vader reneges on his promise to depart with the stormtroopers and uses Han and his shipmates as bait for the Jedi, Luke Skywalker. Having betrayed his friends and then having been double-crossed himself, Lando reevaluates his position. He orders the evacuation of the city and aids in the escape of Han's shipmates, betraying Vader and the Emperor.

This example demonstrates a difficulty in conspiring with malcontents. While conspiracies require recruiting malcontents, these malcontents may be driven by motives such as revenge, a desire to disrupt the status quo, or personal ambition and ego. As a result, conspiring with malcontents exposes the conspirator to unpredictable risks, including betrayal by their own co-conspirators.[3]

Corollary of Maxim 6: Collaborating with malcontents paves the way for betrayal.
Inverse of Maxim 6: The fear of betrayal may itself drive an individual to betray a trust.

Maxim 6, along with its corollaries and inverses, reveals the danger of relying on dissatisfied or unreliable allies. Collaborating with individuals inclined towards disloyalty increases the risk of betrayal. That creates a perilous psychological dynamic: once conspirators suspect one another of impending betrayal, that suspicion alone can drive them to strike first. In seeking to protect themselves, they may preemptively betray their partners, believing that disloyalty offers safety. Yet in doing so, such actions ensure the conspiracy collapses under the crushing weight of mutual distrust.

A person may betray a trust for many reasons. Common motivations include:

- Reward: Seeking personal gain for survival, wealth, or advancement.

- Ideology: Experiencing disillusionment with a political system, bureaucracy, or policy that harms a particular group, culture, gender, religion, or societal role.
- Coercion: Acting under threat of violence, blackmail, or exposure.
- Ego: Driven by a feeling of superiority, resentment, or desire for recognition.
- Sex: Motivated by love, affection, or sexual intimacy.

Greed is often the betrayer's undoing. Spies may be tempted by promises of wealth, politicians by the lure of *graft*, and military industrialists by the prospect of profiteering from war. Each sees opportunity in crisis, exploiting fear and conflict for personal gain while cloaking their actions in the language of duty and security. Yet these ill-gotten rewards often expose them. Sudden affluence, lavish spending, or unexplained fortunes are telltale signs of corruption, marking the first cracks in loyalty and foreshadowing betrayal.

Betrayal can cut both ways. While a subordinate may betray a leader or group, a leader may betray those in their trust. In **The Lost Fleet** universe, the interstellar fleet commander, Admiral John Geary, leads an ambushed, damaged, and demoralized fleet out of enemy-held space and back to friendly territory. Arriving home, Geary's mythical past and near-impossible achievements make him a revered hero to the fleet and the populace. Politicians suspect Geary can act with impunity, even to the point of ousting them and naming himself supreme ruler. And while this might be true, he only desires to serve the political system that he has sworn an oath to protect. But the politicians distrust him. They exploit his loyalty and order him and his fleet on subsequent suicidal missions to eliminate the threat he poses to them. In this case, the politicians legally betray Geary and his command to preserve their tenuous hold on power.

Betrayal can occur anywhere, but it is most diabolical on the battlefield. A surprise attack that betrays a prior alliance, a feigned surrender, or the refusal to offer mercy in the heat of battle may violate a society's laws or norms and often leads to devastating consequences.

Death, destruction, and a spiral of mistrust and retaliation can erupt from a single act of treachery. Those who embrace ruthless tactics, no matter how diabolical, must accept their opponent's escalation to even more diabolical means and ways.

As a result, a society's actions, not its words, reveal its true values and beliefs. Actions provide a window into a society's collective mindset, revealing how it perceives the world and balances competing priorities.

Corollary of Maxims 14 and 17: Actions, not words, reveal how one perceives reality, and inform the world of a society's priorities.

The 2004 reboot of *Battlestar Galactica* depicts an interstellar war between the twelve human colonies and the organic synthetic humans, the Cylons. During their first conflict, the Cylons exploited computer viruses and electronic warfare, prompting humans to abandon sophisticated information technology to prevent further infiltration. After an armistice is signed, the Cylons vanish for over forty years. In that time, the Colonies reinstalled sophisticated information technologies to enhance their capabilities.

Unbeknownst to the humans, the Cylons develop twelve models of organic synthetic humans. A subgroup within the Cylons becomes religious fanatics, viewing humans as an evil force that they must eliminate. They clandestinely embed copies as sleeper agents in key positions throughout human society, aided by Dr. Gaius Baltar, a narcissistic opportunist seduced by a beautiful Cylon posing as a defense contractor. In exchange for intimacy, he grants her access to the Colonies' defense networks. The Cylons begin with subversion, sabotaging defense systems with malicious code to weaken humanity from within.

On the day of the invasion, the Cylon agents disable Colonial defenses, annihilate their fleet, and devastate the Twelve Colonies. Only the vintage *Battlestar Galactica* and its complement of Viper Mark II fighter squadrons, immune to electronic infiltration because of their antiquated technology, escape with a fleet of survivors.

As the human fleet retrogrades, embedded Cylon sleepers within the fleet launch additional subversive operations, including an attempted decapitation of military and political leaders and sabotage of vital resources. Once the survivors realize the Cylons can appear as humans, paranoia sets in, and a "witch-hunt" ensues, further undermining trust and unity. This war is between nuclear-armed conventional military forces, supported and enabled through betrayal and subversion.

DECAPITATION

> "The commander had won the field through the employment of his favorite, practiced strategy: the speartip thrust to tear out the throat... the commander had struck directly at the Overlord and his command coterie, leaving the enemy headless and without direction...Tear out the throat and let the body spasm and die."
>
> — CAPTAIN GARVIEL LOKEN, LUNA WOLVES
> (ABNETT 2006)

Warfare frequently employs means and ways that citizens view as morally repugnant. To navigate this, societies may establish boundaries and rationales for their actions, defining what qualifies as a legitimate military target and acceptable methods for engaging them. Views of permissible violence may vary widely. One society may readily endorse military action to achieve strategic objectives yet reject political assassinations; another may exhibit a far less restrictive approach, imposing no constraints on the use of violence for achieving ends. A society may authorize the pursuit of high-value targets (HVTs), individuals or groups considered *prominent* or *culpable* for violating fundamental principles that justify extraordinary action.

A *decapitation* involves assassinating, kidnapping, incapacitating, or discrediting key individuals through physical, logical, or psychological means. The primary objective of decapitation strategies is to induce strategic paralysis, wherein the removal of pivotal figures degrades the adversary's capacity to operate effectively. Unlike strategies focused on destroying military assets, decapitation seeks to dismantle an enemy's leadership structure, sow chaos, disrupt operations, and erode morale, rendering the adversary unable to act coherently. This strategy may be particularly effective against societies reliant on a single charismatic, autocratic, or religious leader, where the loss of this central figure triggers widespread despair and operational paralysis, significantly hindering their ability to function.

Maxim 53: Cut off the head and let the body die.
Decapitation may be effective in disarming and demoralizing an enemy.

The assassination of Undersecretary Sadavir Errinwright in *The Expanse* is an example of a strategic decapitation strike, removing a key figure whose influence shapes the course of interplanetary war. Errinwright, a senior Earth government official, orchestrates clandestine conflicts and manipulates the protomolecule arms race, pushing humanity towards system-wide war. His death, arranged quietly through political maneuvering and covert action rather than open violence, severs the chain of escalation and destabilizes the aggression he helped engineer. This is a textbook decapitation; removing a high-value leader can cripple command cohesion, halt hostile momentum, and create space for diplomatic recalibration. Unlike battlefield assassinations, Errinwright's removal occurs within the government bureaucracy, proof that in the cold calculus of war, civilians behind a desk can be as decisive a target as a general in the field.

Decapitations produce asymmetric effects. Actors may employ *signature strikes* to remove prominent individuals or *scalpel operations* to eliminate those who possess specific skills or knowledge. The psychological and strategic shock of losing a high-profile target often ripples far beyond the act itself. A single bomb against a key individual can shift the political balance in ways no battlefield skirmish could. Likewise, subversive forces may strike command-and-control networks, critical infrastructures, or strategic weapon systems to destabilize an adversary. In every case, the objective is the same: disable the enemy's ability to coordinate, influence, or retaliate, and force them to respond from a position of confusion and disarray.

In *Star Wars: Return of the Jedi*, during Mon Mothma's strategic briefing to Rebel leadership, she emphasizes that "most important of all, the Emperor himself is personally overseeing the final stages of the construction of this Death Star." This observation is more than mere intelligence. It identifies the Emperor as a critical component of Imperial power, whose presence creates a rare opportunity not merely to destroy a superweapon, but to decapitate the Empire and cripple the system in a single stroke. When power is highly centralized, striking the leader can produce effects far beyond those achievable through attritional warfare.

Decapitation strategies need not rely solely on physical elimination. Subversives may achieve a logical decapitation by targeting the news media and command-and-control systems essential to coordinating civil and military operations. These efforts may range from targeted cyberattacks to broader measures, such as detonating an electromagnetic pulse (EMP) that disrupts communications infrastructure. This approach severs the nerves between the brain and body, disabling coordination without removing the head. Jamming, hacking, or otherwise disabling vital networks can fragment decision-making and cripple an opponent's ability to function as a cohesive force. Logical decapitation offers the added benefit of minimizing physical destruction, enabling a faster post-conflict recovery.

However, leaders must exercise caution when targeting an opponent's capital or leadership infrastructure, because striking the heart of a governing system can produce unintended consequences, including destabilizing the entire society. If the central authority is removed, society may slide into civil war, fall under the control of warlords or criminals, or disintegrate into anarchy. A decapitation strategy may prove ineffective if the targeted society is resilient and capable of replacing its leadership, or when the slain leader becomes a martyr, whose death solidifies opposition and intensifies resistance. A decapitation may also reveal that, while a tyrant's rule ends with their death, a martyr's influence begins with it (Kierkegaard and Dru 2003).

Martyrdom often becomes a potent symbol that motivates resistance against powerful adversaries, turning apparent weakness into strength and inspiring others to resist overwhelming odds. Their self-sacrifice for a cause can unify disparate groups, mobilize the populace, and sustain a campaign against seemingly insurmountable odds.

In *Star Wars*, Luke Skywalker's journey begins with profound loss. Before the Imperial forces arrived on his home planet, Skywalker had no plans to join a rebellion or become a Jedi. His goal was to leave Tatooine and enroll at the Imperial Academy to become an Imperial pilot. He saw the Academy as a way to escape his isolated life and find purpose among the stars. He was restless, frustrated with his uncle's efforts to keep him grounded, and unaware of the true nature of the Empire he hoped to serve.

When Imperial forces arrive, they murder his aunt and uncle, destroy the only home he has ever known, and force him to confront the brutal reality of the Empire. This trauma shatters his sheltered life and propels him into a wider, more dangerous galaxy. In the aftermath, Obi-Wan Kenobi, an aging Jedi Master, becomes Luke's mentor and guide. More than a teacher, Obi-Wan serves as a father figure, offering Luke purpose, discipline, and a connection to a larger legacy.

Obi-Wan's death at the hands of Darth Vader marks another turning point. Witnessing his mentor's sacrifice leaves Luke devastated, but it also deepens his commitment to the Rebel cause. Obi-Wan's deliberate choice to face Vader, and to perish in front of Luke, serves not only as a personal loss but as a symbolic act of defiance. His death passes the torch to the next generation, reinforcing Luke's resolve and providing a rallying moment for the Rebel Alliance. Luke's loss transforms him from a reluctant farm boy into a determined rebel against tyranny.

Those who execute decapitation strikes frequently discover that dealing with the familiar, even if less than ideal, can be preferred to the risk of the unknown. Familiarity allows for a deeper understanding of the other party's strengths, weaknesses, and motivations, enabling more informed decisions about interactions.

An assassination attempt may trigger widespread repression, including mass crackdowns, purges, heightened surveillance, detentions, and violent reprisals that reach far beyond the collaborators. In seeking to reassert control, regimes may target anyone even loosely linked to dissent, further destabilizing society.

Decapitation strategies may be an effective way to cripple a rival by removing a charismatic leader, revered figure, or strategic mastermind. However, these individuals often serve as the ideological anchor, a unifying force, or the tactical brain of a movement. Their removal can fracture command structures, sap morale, and trigger infighting among successors. The group may fragment, weaken, or scatter, and without a clear head to follow or negotiate with, the remaining elements may mutate or radicalize, making them harder to locate and neutralize

In *The Expanse*, the death of Marco Inaros results in the Free Navy losing its central figure, the one who united the Outer Planets under a banner of violent rebellion. Marco's charisma, tactical aggression, and populist rhetoric gave the Belt a rallying voice. Once he is gone, the Free Navy fractures. Some factions lay down arms, others flee, and a few continue to resist, without unity or clear command. Losing Marco weakens the movement's momentum and opens the door for political reconciliation. Yet, it also disperses radicalized remnants, who operate independently and are harder to monitor or suppress. This illustrates both the power and the peril of decapitation as a strategy.

Striking an opponent's leadership or severing command-and-control risks other unintended effects. When leaders lose the ability to direct their forces, others may misinterpret ongoing operations as a sign of further aggression or an unwillingness to surrender. This could inadvertently prolong a conflict or trigger an escalation.

In *The Lost Fleet* series, the majority of the Alliance interstellar fleet ventures into the territory of their century-long adversary, the Syndicate, anticipating a decisive battle. Deep within enemy territory, the fleet discovers that it has entered a trap. Facing imminent destruction, senior Alliance commanders disembark under a truce to negotiate with the Syndicates. The Syndics betray this trust and assassinate the Alliance leaders, plunging the fleet into chaos far behind enemy lines.

John "Black Jack" Geary emerges as the new leader of the battered and demoralized fleet, rallying them to escape immediate peril. He leverages the Syndicate's betrayal and brutality to strengthen the survivors' resolve as they retreat to friendly territory. This traumatic experience teaches survivors a harsh lesson: they can never trust the Syndics in future negotiations (Campbell 2006).

Leaders should take care that decapitation strategies do not spiral into attacks on civilians. When violence targets non-combatants to instill fear and force political or ideological change, the action crosses into *terrorism*. Terrorism is a form of subversive warfare defined not by its ends, but by its ways; it is not an objective, but a brutal way of exerting pressure. Such tactics are often employed by combatants who lack the military strength for direct confrontation and who instead use fear and chaos to influence public perception, provoke overreaction, or extract political concessions.

Corollary of Maxim 14: One person's terrorist is another person's freedom fighter.[4]

Violence against the defenseless evokes profound shock and horror. Such actions often distance terrorists from mainstream society, fuel retribution against both the perpetrators and their supporters, and increase the cost and effort to restore society after hostilities end. Most importantly, each terrorist attack signals the terrorist's willingness to disregard the innocent and treat violence against non-combatants as a legitimate tool. If such actors gain power, there is little reason to expect their principles or methods to change. Though terror may achieve short-term results, it ultimately undermines its own legitimacy and becomes a self-defeating strategy.

Corollary of Maxims 14, 19, 21, and 26: Terror may briefly bend wills, but it destroys trust and prevents lasting resolution.

LIFE ON THE RUN

Guerrillas face a constant struggle. They typically lack the conventional firepower needed to confront their foes directly, which prevents them from controlling a defended region where they can rest, grow crops, or establish large supply depots. Even training is difficult because it may alert the enemy to their presence. Traditional defensive tactics, such as fixed fortifications and static support bases, become liabilities against an opponent capable of maneuvering freely and applying overwhelming force. If guerrillas adopt defensive positions, their opponents need only trap them in a decisive engagement to destroy them.

Corollary of Maxim 10: Defensive positions are a vulnerability to the subversive.

The subversives' critical component is their support system, making logistics both their strength and their weakness. A conventional military force receives its support from its state. Subversives often obtain support, willingly or through coercion, from a portion of the population. They may use intimidation or exploit commonalities with supporters, such as culture, religion, race, or political beliefs. Or, guerrillas may buy support, provided they have something to barter.

When subversives operate in their native land, the populace becomes a double-edged sword. A supportive populace may provide vital logistics, safe havens, and near invisibility. However, these benefits hinge on maintaining the public's trust. Accidental or deliberate civilian casualties can shatter this fragile trust, eliminating their support and transforming allies into informants and collaborators. This not only weakens the insurgency but also hands the enemy a potent propaganda tool. By highlighting civilian suffering, the enemy can frame the subversives as reckless terrorists, further eroding their legitimacy and public sympathy.

Operating in a foreign land presents even greater challenges for subversives. The local population often views outsiders with suspicion, making it difficult for them to build support networks or secure safe havens. Cultural differences and language barriers further complicate communication, coordination, and integration with local resistance groups.

The enemy is often the immediate and primary source of a guerrilla's weapons and ammunition. Scavenging their opponent's weapons and supplies after engagements, or raiding their convoys, bases, and warehouses, may be an effective way for subversives to resupply and grow military power. Subversives in this situation must rely on meticulously planned operations and a focus on smash-and-grab raids.

Sometimes it can be advantageous for subversives to receive support from external third parties. However, as Maxim 6 cautions, as subversives become beholden for support, refuge, and intelligence, they create dependencies, and since every dependence is a vulnerability, those who provide support often gain control over subversives.

A subversive's survival relies on deception and surprise, primarily by denying opponents intelligence about their members and operations. Maxim 36 cautions that only the ignorant and the dead will maintain secrets. As a movement grows, it becomes increasingly vulnerable to infiltration by spies, double agents, or betrayal from disillusioned or coerced members. To mitigate these risks, subversives often decentralize command and control, logistics, and combat operations into small, independent cells that form a distributed network rather than a traditional hierarchy. Each cell operates semi-autonomously, with members knowing only those within their group and, at most, a few trusted contacts in others, typically limited to designated leaders or family ties. These compartmentalized cells may combine to conduct an operation and then dissolve back into the terrain or populace. This compartmentalization ensures that no individual holds the full picture of the organization, preserving the broader movement even if a single cell is exposed.

The Rebel Alliance in **Star Wars** is a subversive group that operates through decentralized cells. The Alliance is made up of many groups and individuals united in their opposition to the Galactic Empire. The Alliance's cells are initially small and independent, making it difficult for the Empire to infiltrate the Alliance and discover its overall plans.

Operating undercover imposes relentless psychological strain. Guerrillas live in a constant state of vigilance, unable to rest, while their loved ones face the threat of persecution, and essential supplies such as food, ammunition, and medicine remain scarce. Capture often results in imprisonment, torture, or execution.

Prolonged guerrilla conflicts inflict severe hardship on the populace and economy, especially in urban areas where collateral damage harms civilians and disrupts society. Fearing popular support for the insurgency, the incumbent power may resort to brutal tactics, even using weapons of mass destruction against suspected collaborators. In a guerrilla war, everyone becomes a target, and everyone suffers.

The success of subversives depends on their ability to integrate with local resistance or ignite a broad popular uprising, both of which require a unified strategy. A consistent and cohesive effort conserves limited resources, amplifies tactical impact, and conveys a clear message to the population. Disjointed or rogue actions, especially those involving brutality, can alienate potential supporters, erode legitimacy, and turn the population against the cause.

A subversion campaign is a protracted war of exhaustion. Success hinges not only on the persistence of the subversives but also on the commitment of their support base. If their opponent doubts this resolve, they may wait them out, allowing the movement to wither and collapse on its own. The drawn-out nature and brutality of these struggles exact a heavy toll. For most subversives, this path is not one of glory but of desperate necessity, borne from hardship rather than choice.

COUNTER STRATEGIES

"There's a reason you separate the military and the police. One fights the enemies of the state. The other serves and protects the people. When the military becomes both, then the enemies of the state tend to become the people."

— WILLIAM ADAMA, ADMIRAL, BATTLESTAR
GALACTICA (GRABIAK 2005)

Military leaders often default to the most direct approach to counter subversives: track, trap, and destroy. However, attacking guerrillas who can blend into the environment or populace is attacking their strength, often turning counter-guerrilla operations into long-term, bloody, and frustrating campaigns. Alternatively, if an opposing force is willing and able to destroy that population along with the guerrillas' support network, concealment loses its value, but possibly at great moral and political cost.

The *Star Wars* universe illustrates the dangers of using heavy-handed tactics against a rebellion. The Empire's brutal repression back-fires, inadvertently creating martyrs like Obi-Wan Kenobi, Saw Gerrera, Jyn Erso, Cassian Andor, and the entire population of the destroyed planet of Alderaan. These deaths fuel the flames of insurgency, swelling the ranks of the rebellion. This highlights the peril of a purely military response to a subversive movement: it might strengthen the resolve of those fighting for change.

A more sustainable and effective approach may involve government forces employing social campaigns, deception to delegitimize guerrillas, dismantling networks by severing connections between cells, or pursuing political solutions that address grievances and the root causes of the violence. These *indirect approaches* exploit guerrilla vulnerabilities by targeting the foundations of their support, such as poverty, oppression, or social injustice, weakening the insurgents' appeal and eroding their popular support. Psychological operations may aim to win the populace's "hearts and minds" by exposing the evils of the rebellion and presenting the ruling authority as a more capable and just alternative. However, these efforts can backfire if the local population perceives them as insincere, foreign, or offensive, or as attempts to eliminate their customs.

Maxims 19 and 26 offer opposing views on subjugation: Maxim 19 suggests that aggressive military action may compel obedience, while Maxim 26 cautions that violence breeds resentment and hatred. Consequently, internal security forces often prioritize deterring subversive behavior over identifying and prosecuting specific subversive activities. As explored in Chapter Nineteen, deterrence depends on the perception of a credible threat. To deter guerrillas and their supporters, security forces must maintain a visible presence, yet that visibility, such as troops patrolling a city, often backfires by fueling resentment, unrest, and civilian suffering (Hart 1954). This dynamic benefits subversive forces, as Maxim 49 advises them to avoid a fight until conditions favor them. Since security forces must remain observable to maintain credibility, subversives can remain concealed and strike only when the circumstances favor them.

Countering an insurgency may require the government to publicly uphold its civilization's moral principles while quietly adopting selected insurgents' tactics. Counter-guerrilla operations can wear down conventional troops, who grow frustrated by chasing elusive opponents and by dealing with a populace that often sympathizes with them. Unable to distinguish friend from foe, troops may lash out or adopt punitive measures, worsening relations with civilians and ultimately helping the subversives. Effective counterinsurgency, therefore, demands persistent intelligence, surveillance, and targeting; absent those elements, its efforts become ineffective or counterproductive. Once counterinsurgent forces identify the likely locations and timing of guerrilla activity, success depends on acting with speed, surprise, and sustained pressure to fix, trap, and destroy the insurgent force before it can disperse. Conventional operations often fail against elusive, decentralized opponents, so leaders may need to empower junior military commanders who can think like guerrillas and employ deception, ambushes, and small-unit operations to locate, isolate, and destroy them.

Targeting guerrilla logistics, intelligence, and support networks can be more effective than massed force. Denying insurgents access to local resources, to supplies from foreign sponsors, or even to their families may weaken their resolve. Governments might relocate populations to refugee or detention camps, or, more brutally, eliminate them to sever assistance to the insurgency. Such measures may drive civilians to turn against the guerrillas out of resentment or desperation. Blockades and interdictions can isolate subversive forces while framing the inconvenience as the cost of harboring insurgents.

Yet governments often discover that, while military force can obliterate people and infrastructure, it cannot eliminate an idea unless every believer has been extinguished.

Corollary of Maxim 57: Military force may defeat force, but ideas endure until the last believer is eliminated.

SPECIAL FORCES

"From this moment until we win this war, the only easy day is yesterday."

— SERGEANT MAJOR FRANK BOUGUS, UNITED STATES MARINE CORPS, SPACE AVIATOR RECRUIT DEPOT (MORGAN AND WONG 1995)

Societies may form paramilitary units or specially trained military forces, sometimes referred to as special forces or commandos, to defeat enemy insurgents or to operate as insurgents themselves to undermine an opponent's government or military. These small, elite teams can offer a cost-effective alternative to large-scale conventional military forces and provide leaders with plausible deniability and reduced political risks.

Special forces troopers are not merely elite conventional forces; they possess special skills, tactics, training, and equipment tailored to complete complex and dangerous unconventional missions. Their special skills may include fluency in foreign languages, cultural knowledge, and training to operate with foreign fighters. Commandos are often capable, loyal, and trusted to work independently in small groups and on sensitive projects.

Their missions may include:

- Unconventional Warfare: Supporting foreign guerrilla forces to resist, disrupt, or overthrow a foreign government, offering a low-risk means of subversion.
- Foreign Internal Defense: Assisting a foreign government in countering internal threats.
- Special Reconnaissance: Gathering intelligence in hostile environments while avoiding detection to aid allies or mislead adversaries.
- Direct Action: Conducting raids, ambushes, sabotages, manhunts, and assassinations.
- Infrastructure and Police Support: Suppressing riots, assisting emergency services, and inspecting for contraband.

The United Nations Space Command (UNSC) in *Halo* relies heavily on its Office of Naval Intelligence (ONI) to manage strategic deception and unconventional warfare. The ONI developed the Spartan Program, originally to suppress insurgencies in the outer colonies. As rebellion spreads across distant worlds, the UNSC turns to the Spartans, physically augmented and psychologically conditioned soldiers taken from childhood to perform missions too dangerous or politically sensitive for conventional forces. Their early operations focused on counterinsurgency, targeted eliminations, and intimidation campaigns designed to weaken separatist resolve.

When war erupts with the alien Covenant, the Spartans are redeployed to face a far more powerful enemy. Their mission shifts from suppressing internal unrest to conducting irregular warfare against a technologically superior force. Through stealth, sabotage, and precision strikes, Spartans disrupt Covenant logistics, eliminate high-value targets, and gather battlefield intelligence. This transformation highlights the paradox of its creation. Soldiers once trained to enforce control within human territory become vital to humanity's survival by adopting the same subversive tactics they were meant to eliminate (Futamura 2010; Mimica-Gezzan et al. 2015; Nylund 2001).

While special forces units are typically in exceptional physical condition and receive extensive training, their most important asset is their mindset. Selection and training emphasize adaptability and creative problem-solving under pressure. This demands the mental toughness to commit acts of violence when necessary, a "can-do" attitude that ensures mission completion despite adversity, and discipline to protect state secrets.

Maxim 10 warns to safeguard opportunities. Commandos are costly to train and are a powerful force multiplier. They should not be used as conventional military forces.

CONCLUSION

Both sides in a subversive war must demonstrate persistence because if their resolve falters, their opponent will outlast them. One side may employ sound tactics and win battles yet still lose the war if those victories come at the cost of dwindling military strength (attrition) or eroding popular support (exhaustion). Over time, political realities may force it to capitulate, or the society may become too weak to continue and face defeat. A leader must maintain awareness of the overall picture, anticipate unintended consequences, and identify tactical issues that create adverse strategic effects.

When subversives dismantle their own society's political and military institutions in an attempt to take control, they may inadvertently expose it to invasion by a third party. This may have been the third party's true aim all along, not to support the revolt, but to engineer the society's collapse. In such cases, the third party employs a bait-and-bleed strategy, manipulating rivals into a prolonged, mutually exhausting conflict. Once both sides are sufficiently weakened, it can intervene and eliminate both.

On the other hand, nothing unites a society faster than a common external threat. In *Dune*, the Emperor perceives House Atreides as a rising power that endangers his rule. Rather than confronting them directly, he adopts an indirect approach. After eighty years of ruthless exploitation by House Harkonnen, the Emperor transfers stewardship of Arrakis, the galaxy's only source of melange, an invaluable spice that prolongs life and enables interstellar navigation, to House Atreides. This imperial decree compels the Atreides to abandon their defensible home world and take possession of the Harkonnens' holdings, an act that humiliates the Harkonnens and provokes their fury. The Emperor pits two powerful rivals against each other, disguising his manipulation as an inter-house feud. Predictably, the Harkonnens strike first, annihilating House Atreides and fulfilling the Emperor's objective without revealing his hand.

As is often the case, treachery proves to be its own undoing. The heir to the Atreides dynasty unites the surviving members of his house and aligns them with the formidable native inhabitants of Arrakis. Together, they wage a guerrilla war, first toppling the Harkonnens and then the Emperor himself (Herbert 1965).

CHAPTER 21

POLITICAL STRATEGIES

"…a good military commander spends the lives of their people reluctantly and with regret but does spend them when necessary. The good politician does the same thing with principles. There aren't any fine burials for sacrificed principles, though."

— VICTORIA RIONE, CO-PRESIDENT OF THE CALLAS
REPUBLIC AND ALLIANCE SENATOR (CAMPBELL
2011)

Military force and diplomacy represent two fundamental means by which a society interacts with the galactic stage. The military stands as the embodiment of force, wielded in times of war, most often to defend against external threats or achieve societal objectives. Diplomacy is the instrument of peace, fostering dialogue, cooperation, and resolving conflicts through negotiation.

Although they serve the same society, political leaders and military officers confront problems through fundamentally different decision-making lenses. Military leaders charged with the physical protection of their society tend to view the world in absolutes. Their decisions stem from a clear-cut sense of right and wrong, grounded in what they perceive as the best interest of their society. In this worldview, there is little room for ambiguity.

Politicians cannot afford such rigid thinking. Their role requires compromise, balancing limited resources, forging alliances, and advancing competing interests. To military professionals, these decisions may seem subjective or self-serving. Politicians may reduce force size to satisfy fiscal constraints, approve procurement contracts that favor domestic industries over combat readiness, or deploy forces to secure trade routes and natural resources that seem only loosely connected to national defense.

At times, politicians might initiate or prolong a conflict to secure their party's dominance. Although critics may view this as self-serving, the politicians themselves may believe that only their leadership will ensure the state's survival. This is depicted in Orwell's *Nineteen Eighty-Four*, where Oceania's ruling party perpetuates endless war to preserve its hold on power.

The contrast between warriors and politicians becomes more pronounced at a war's end. Political leaders charged with concluding a brutal conflict may find themselves negotiating with the very enemies who initiated it and bear responsibility for the death of warriors and citizens alike. For military personnel steeped in a culture of duty and sacrifice, this diplomacy, regarded by statesmen as a necessary compromise, may be perceived as surrender, betrayal, or even outright treason.

POLITICAL POWER

> "And don't forget—she's a politician. They're not to be trusted."
>
> — OBI-WAN KENOBI, JEDI MASTER (LUCAS 2002)

A politician's ability to achieve objectives and inspire others often depends on their personal power, charisma, and popularity. Successful political leadership, particularly among humans, requires the following essential skills:

- Unity of effort: Align diplomatic actions with strategic ends, objectives, and interests to ensure consistency and effectiveness.
- Negotiation: Employ persuasion to shape outcomes, avoid crises, and pursue mutually acceptable solutions. A breakdown in diplomacy always leads to a crisis and may result in violence.
- Understanding: Gather and interpret information to comprehend others' values and intentions. Leaders must excel at *perspective-taking* by grasping political, military, economic, and technical factors, anticipating how others pursue their interests, predicting how they will perceive and react to those interests, and shaping decision-making accordingly.
- Influence: Shape the behavior of leaders and societies through messaging, incentives, and strategic positioning, including the careful application of timing, pressure, and diplomatic maneuver to advance objectives and avoid unnecessary confrontation.
- Trust: Maintain credibility. Diplomats known for deception or bad faith forfeit future negotiating power.
- Clarity: Communicate views and ideas effectively across differing interests, cultures, and languages, applying precision or ambiguity as circumstances require.
- Pragmatism: Present viable proposals supported by genuine authority, resources, and intent, or risk being dismissed as irrelevant.
- Initiative: Leaders must proactively pursue an agenda or risk being overtaken by those who do.

It is important to note that these traits reflect effective human political behavior shaped by historical experience. They may not apply when interacting with radically different cultures.

Paul Atreides in **Dune** is both a political and military leader, and an illustration of how political power is created, consolidated, and legitimized through personal authority, perception, and strategic behavior.

His rise from the son of an assassinated Duke to Emperor of the known universe demonstrates that a politician's ability to achieve objectives and inspire loyalty often depends on the deliberate cultivation of charisma, credibility, and perceived inevitability.

Paul aligns his objectives with an end, and thereby unifies the fragmented Fremen under a single political and military vision. Through careful negotiation, he reshapes traditional customs without shattering social cohesion, proving that diplomacy, when skillfully applied, can prevent violence while strengthening authority. His deep understanding of Fremen psychology, religious belief, and cultural expectation allows him to position himself not merely as a leader but as the fulfillment of prophecy, transforming personal power into political legitimacy.

His mastery of intelligence, both formal and intuitive, enables him to anticipate the intentions of the Emperor, the Harkonnens, and the Landsraad, the governing body representing the Great Houses. Paul exerts influence through symbolism, controlled narrative, and carefully staged demonstrations of power, shaping how both followers and adversaries perceive him. By cultivating trust among the Fremen while projecting fear and inevitability toward his opponents, he strengthens his bargaining position and consolidates his control.

Paul's communications alternate between precision and ambiguity to reinforce authority while maintaining flexibility. His pragmatism is evident in his willingness to balance idealism with survival, preserving military strength and political momentum even when moral compromises are required. Through maneuver and initiative, he forces events to unfold on his terms, rarely allowing opponents the luxury of setting conditions.

Paul's ascent illustrates that political power does not emerge solely from position, but from the fusion of personal authority, cultural alignment, strategic foresight, and controlled narrative. His case also reveals the inherent danger of such power, as the same traits that create unity and stability may also unleash forces beyond the leader's control.

Just as military operations frequently serve political objectives, many political endeavors carry implicit military consequences. Likewise, strategies such as divide-and-conquer, manipulation of belief systems, and strategic coercion may operate as effectively in political arenas as they do on the battlefield.

Next, we address a few political strategies relevant to warfare.

ALLIANCES

"Brother, this peace, this alliance, is like a living death to warriors like us."

— KORRIS, CAPTAIN, KLINGON DEFENSE FORCE
(BOWMAN 1988)

A society may negotiate an alliance with other parties for many reasons, including trade, political actions, or military support. The goal of forming or joining an alliance is to make a society stronger or increase its security.

A society that has sufficient power to unilaterally defeat an opponent may still ally with a third party to conserve its resources. Or a civilization may seek alliances to gain a military or commercial advantage, such as access to commercial ports or markets.

The most common military alliances include the defense pact, a mutual agreement in which members agree to defend one another against external threats, often adopted by weaker societies to deter or counter stronger adversaries, and the non-aggression pact, a reciprocal promise not to attack, typically employed to secure neutrality during rising tensions or impending conflicts.[1]

While alliances are intended to support a society, they may unintentionally hinder or harm it. If societies fear that a neighbor may resort to violence, they might try to preserve the peace by creating a defense pact. However, an alliance that obligates members to respond to aggression can lead to *chain-ganging*, where a threat against one member compels others to act. In this case, a *security dilemma* may arise, in which a minor skirmish unintentionally escalates into a much larger conflict as each society honors its commitments. In such cases, alliances may strip their members of initiative, forcing them into a conflict few might not have chosen on their own.[2]

A powerful society must exercise utmost caution when forging ties with a weaker society. The demands of the weaker entity may sap the stronger entity's strength, eroding its initiative and forcing it to respond to unreasonable requests. This entanglement can lead to significant harm to a powerful society.

Likewise, within an alliance, a member may "pass the buck" by manipulating other members to undertake costly actions while it preserves its own resources and strengths. Such behavior may involve encouraging an ally to deter or engage a third party, allowing the instigating state to conserve its strength for future strategic advantage.

Maxim 54: An alliance is a double-edged sword.
Alliances amplify collective power while constraining autonomy
and inviting exploitation.

An alliance may be *overt* or *covert*, with its terms either openly declared or deliberately concealed. The discovery that a neighbor has aligned with an unfavorable faction can unintentionally provoke suspicion, counter-alliances, or even conspiracy against that society. Likewise, dependence on foreign weapons, fuel, or equipment grants the supplier considerable influence over the recipient's economy, society, and political mechanisms.

A society may also place undue faith in the written promises of its treaties, such as pledges to aid a neighbor in wartime, only to learn that its ally no longer considers honoring those obligations to be in its best interest. Here, the alliance creates a false sense of security and encourages flawed assumptions during planning. The persistent risk of betrayal may therefore compel a society to monitor its allies through intelligence collection, seeking reassurance that they intend to fulfill the terms of their agreements.

Corollary of Maxims 37 and 54: Relying on goodwill is perilous.
Unenforceable treaties are mere promises on paper.

In the *Star Trek* universe, the Khitomer Accords establish an alliance between the United Federation of Planets and the Klingon Empire, two powers historically bound by hostility. The agreement creates a defense pact designed to deter aggression and stabilize the quadrant. For the Federation, it preserves resources and extends strategic reach. For the Klingons, it provides legitimacy and political leverage. Yet the alliance repeatedly demonstrates its double-edged nature. Internal Klingon political instability, coupled with mutual suspicion, forces both parties into actions driven by obligation rather than choice. During events such as the Klingon civil war and later conflicts in Deep Space Nine, each side conducts surveillance on the other, anticipates betrayal, and manipulates the alliance to protect its own interests. Even the Federation, despite its superior power, finds itself reacting to Klingon demands to avoid the collapse of the accord. The Khitomer Accords illustrate how alliances amplify collective strength while simultaneously eroding autonomy, fostering mistrust, and drawing members into conflicts they might not otherwise have chosen.

ISOLATION

Isolation is a strategy of avoiding political, military, or economic commitments and refraining from involvement in others' affairs. It emphasizes self-reliance and the advancement of one's own society by rejecting alliances and foreign entanglements. Groups may insulate themselves by eliminating external dependencies, believing this frees them from the need to maintain a large, expensive military.

Following a prolonged or costly war, a society may turn to isolation to avoid repeating past entanglements. One approach is to reject political and military alliances while continuing to engage in trade and travel. However, this approach overlooks the interdependence of power, as economic activity inevitably influences political and military affairs.

Isolationism is a passive strategy. Though often seen as risk-averse, it cedes the initiative to others who may seize emerging opportunities. This passivity can create a power vacuum that opportunists may fill with their own political, military, or economic influence. Over time, a society that refuses to assist others may find itself abandoned when it faces its own crisis.

Inverse of Maxim 13: Pursuing war only as a last resort cedes initiative to an aggressor.

A conservative society intent on preserving the status quo may view change as evil and embrace isolationism. However, as noted in Chapter Fourteen, social and technical evolution continues, regardless. By halting its own progress while others advance, a society risks *strategic obsolescence*.

An extreme form of isolationism avoids all involvement in external affairs. However, others may perceive neutrality as implicit support for an incumbent power or silent approval of emerging aggression. As a result, inaction is rarely viewed as genuine impartiality.

In the *Foundation* universe, as the Galactic Empire begins to collapse under the weight of internal decay, political stagnation, and failing institutions, many outlying systems choose to isolate themselves from Imperial authority to preserve autonomy and avoid being dragged down by a dying power. These peripheral worlds sever political ties, withdraw from shared defense structures, and reject Imperial oversight, believing that distance and neutrality will shield them from the chaos spreading outward from the dying Empire's core.

In practice, this isolation proves disastrous. Without the deterrent force of Imperial fleets or the stability provided by interstellar alliances, these systems become exposed and vulnerable. Fragmenting warlords, opportunistic regional powers, and emerging petty empires exploit their lack of protection. Lacking coordination, shared intelligence, or collective defense, the isolated systems are absorbed, conquered, or coerced into submission. What was intended as a strategy of preservation becomes a pathway to subjugation, demonstrating that withdrawal from a collapsing order can leave a society defenseless in the very moment when cooperation is most essential.

NEUTRALITY

"Being neutral does not make you innocent."

— JONATHAN ARCHER, CAPTAIN, USS ENTERPRISE
(NX-01) (CONTNER AND JACQUEMETTON 2002)

Related to isolationism is *neutrality*, in which a society does not align with either side in a conflict. Prior to hostilities, a country may declare its *non-alignment* using pacts or public announcements. By not siding with either party, they may hope to avoid being drawn into a conflict.

Maxim 55: Beware of binary choices.
The world is not binary, but war often provokes binary decision-making: "You're either for me or against me."

Neutrality's promise of safety is not guaranteed. Declaring oneself a neutral party to a conflict doesn't guarantee that a society will not be attacked by the belligerents. Like all military and political concepts, a willingness to recognize neutrality is determined by each party's philosophy. For some, honoring neutrality is an inviolable rule or law, and for others, an inconceivable constraint on political and military ways.

The harsh reality is that the weak may declare themselves neutral, only to be subjugated or conquered by the strong. This is because neutrality may be viewed as weakness, and isolation may further weaken a society's position. This reflects the power dynamics and brutal consequences of Maxim 1.

Corollary of Maxim 1: The weak aren't neutral; they're conquered.

Both sides in a conflict may view neutrality as complicity, interpreting a refusal to fight for change as tacit support for the status quo. In **Rogue One: A Star Wars Story**, rebel extremist Saw Gerrera challenges Jyn Erso's indifference toward the Empire. When he asks if she can stand to see the Imperial flag reign across the galaxy, she replies,

"It's not a problem if you don't look up" (Edwards 2016). The irony is that those who try to avoid taking a side often find themselves crushed between the two.

**Inverse of Maxim 13: Not deciding is a decision.
Anything you ignore, you accept.**

Declared neutrality may postpone direct involvement, yet it can backfire. Belligerents may come to suspect the neutral party of harboring intelligence agents or covertly supporting their adversaries. Moreover, neutral territory can present a strategically convenient avenue for invasion or occupation, undermining the very security neutrality is meant to preserve.

Consider a scenario in which Party A builds an invasion force threatening Party B. Party B might view a neutral neighbor as a vital buffer, allowing it to concentrate its limited military resources on defending against Party A. As a result, Party B assumes its powerful defenses will deter aggression. Yet a critical question emerges: if Party A is already willing to violate sovereign borders and risk condemnation, what would stop it from violating neutrality as well?

From Party A's perspective, the neutral party might present a convenient invasion corridor, bypassing Party B's defenses and granting easier access to its opponent's territory. In such a scenario, the dependency on neutrality offered a false sense of security while simultaneously creating a vulnerability.[3]

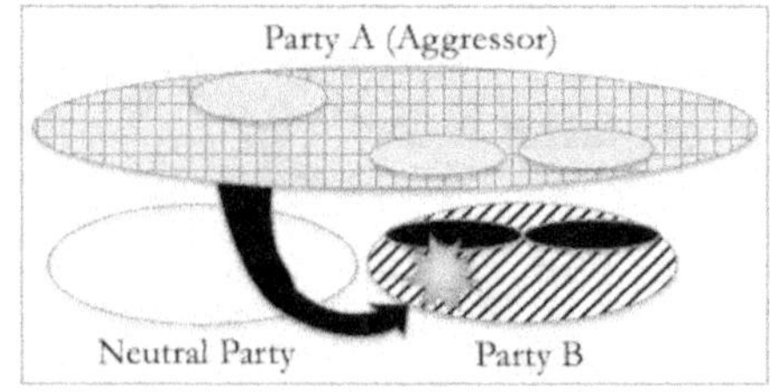

Party A violates a neutral region to bypass Party B's defenses.

Queen Amidala's declaration in *Star Wars: The Phantom Menace,* "I will not condone a course of action that will lead us to war," captures the noble but flawed logic of neutrality in the face of aggression (Lucas 1999). Though intended to protect her people from violence, her reluctance to act ceded the Trade Federation the initiative, enabling them to invade and occupy Naboo unopposed. In military strategy, neutrality in the presence of open hostility concedes the advantage to the aggressor, granting them time and freedom to act without consequence. By hesitating, a leader may forfeit the opportunity to shape the battlespace or rally allies, leaving their society vulnerable to the ambitions of more assertive and ruthless actors. Amidala's delay nearly cost her planet its sovereignty, demonstrating that inaction pursued in the name of peace can unintentionally hasten defeat.

INTERVENTION

"He who controls the spice controls the universe."

— BARON VLADIMIR HARKONNEN, RULER OF
HOUSE HARKONNEN (HERBERT 1965)

Intervention refers to the use of any source of power short of open warfare to influence another society's forms of power. A society intervenes in others' activities for various reasons, including commerce, colonization, expansion, exploration, or ideological promotion. A powerful society interferes in another's affairs when it believes it possesses sufficient power or influence to support its actions and when the expected benefits outweigh the potential costs. Conversely, a weaker society may intervene in the affairs of a stronger one to influence outcomes indirectly while avoiding confrontation. Such interventions may be perceived as acts of salvation by some and of intrusion by others.

Once a society deems the status quo untenable, it may operate along a spectrum ranging from peaceful negotiation and mutually beneficial trade to coercive techniques such as sanctions, embargoes, or trade agreements intended to shape its opponent's behavior. Intervention may also take informational or ideological forms, including news media manipulation, propaganda, the orchestration of protests, the promotion or discrediting of belief systems, or the formation of alliances to secure third-party pressure.

At its most aggressive, intervention may involve efforts to undermine or overthrow a neighboring regime through *covert* or *clandestine* military operations, including raids, piracy, and subversive activities aimed at achieving a desired end without triggering open warfare. These actions transport the hardships of war onto the opponent's territory, shielding the intervening society's own populace from the direct costs of war.

If discovered, such actions may trigger a *tit-for-tat* dynamic in which each party retaliates or escalates until one yields, or the situation degrades into open warfare.

In **Star Trek: The Next Generation**, the Klingon Civil War is an example of intervention short of open warfare. When the Federation discovers that the Romulans are covertly supplying the Duras faction to install a more compliant Klingon leader, it chooses not to engage the Klingons directly. Instead, the Federation intervenes through indirect means, deploying a tachyon detection network along the Klingon border to expose and neutralize Romulan support without firing a shot. This action alters the balance of power within the Klingon Empire while allowing the Federation to avoid direct conflict and preserve the appearance of neutrality.

To some Klingons, this interference is an alliance saving the Empire from foreign manipulation. To others, it is an unacceptable intrusion into sovereign affairs. This demonstrates how a powerful society may intervene through technological, informational, and strategic pressure to shape outcomes, redirect political authority, and prevent an unfavorable regime from emerging, all while remaining below the threshold of open war. In doing so, the Federation shields its population from the immediate costs of conflict while still imposing decisive influence over another civilization's political future.

CONTAINMENT

> "I will not sacrifice the Enterprise. We've made too many compromises already, too many retreats. They invade our space and we fall back. They assimilate entire worlds and we fall back. Not again. The line must be drawn here. This far, no further."
>
> — JEAN-LUC PICARD, CAPTAIN, USS ENTERPRISE (NCC-1701-D) (BERMAN, BRAGA, MOORE, AND FRAKES 2009)

A containment strategy seeks to compel a society to cease expansion or refrain from pursuing it. Fundamentally, it is a defensive strategy designed not to invade an adversary but to prevent that adversary from gaining additional territory, resources, or influence.

Containment strategies frequently rely on overt measures such as political alliances and economic sanctions to restrain a perceived threat. They may also include conspicuous displays of military force intended to deter or intimidate; a method sometimes referred to as gunboat diplomacy. Alternatively, containment can take covert forms, employing financial manipulation, ideological influence, and alliance-building to influence behavior without direct confrontation.

While containment may or may not involve combat operations, it is always confrontational. The party employing such a strategy must appear to possess sufficient military power to back up its demands.

Effective containment depends on understanding an opponent's leadership, motivations, and societal dynamics. A public show of force intended to deter a dictator may instead provoke an escalation if perceived as humiliation. In contrast, negotiations offering a dignified means to end a conflict may achieve objectives without bloodshed. To counter an ambitious aggressor, societies must form alliances early, before the aggressor consolidates power by conquering weaker neighbors. Delay risks allowing the aggressor to reach a tipping point where future resistance becomes futile.

Corollary of Maxim 25: Use alliances to contain a strong or aggressive opponent.

In the *Star Trek* universe, the Borg represent a case for a containment strategy. The Federation recognizes that the Borg cannot be negotiated with, deterred through goodwill, or integrated through diplomacy. Their defining imperative is perpetual expansion through assimilation, making any form of accommodation equivalent to surrender. As a result, Starfleet adopts a posture focused on halting further Borg encroachment rather than attempting conquest of Borg space. Defensive lines are drawn, rapid response protocols are developed, and specialized tactics such as targeted strikes, technological countermeasures, and strategic withdrawal are employed to prevent the Borg from gaining additional territory, knowledge, or access to core Federation systems.

Recognizing that no single power can withstand the Borg alone, the Federation forges critical alliances with former rivals, including the Klingons and Romulans, coordinating intelligence, fleet operations, and technology sharing to strengthen collective resistance. The objective is not the destruction of the Borg as a civilization, but the denial of further expansion. This illustrates containment in its purest form: a defensive strategy that accepts the adversary's existence while committing fully to preventing its growth, influence, and reach.

CONCLUSION

The intricate dance of political strategy reveals a fundamental paradox: peace sometimes demands the same instruments that can lead to war. While isolationism and neutrality may offer a tempting escape, the reality of the political battlespace often requires a more nuanced approach. Leaders must continually respond to shifting threats and opportunities, sometimes forming reluctant alliances in the face of overwhelming danger.

In *Star Trek*, for instance, long-standing adversaries such as the Federation, the Romulans, and the Klingons unite against the Borg. This coalition illustrates how necessity can transcend ideology, compelling cooperation where distrust once prevailed. Yet power in all its forms remains a double-edged sword. Efforts to preserve peace or stability can produce unforeseen consequences. In *Star Trek: The Next Generation*, Starfleet's involvement in the Klingon Civil War, undertaken to prevent Romulan interference and preserve the balance of power, nearly triggers open conflict and sows enduring resentment among Klingon factions. What begins as a stabilizing action can swiftly spiral into chaos, demonstrating that even well-meaning strategies may ignite the very instability they seek to avert.

Strategies are rarely static. As government policies evolve, military priorities often follow. When a military leader's convictions or actions diverge from their society's shifting political aims, tensions arise between the armed forces and the state they serve. This tension can lead to internal struggles, not only over the methods and conduct of war, but over its very purpose and ends, and the legitimacy of the war itself.

Moreover, when this divergence deepens, the military risks being seen as insubordinate, or even as an enemy of the state. Their loyalty may be questioned, their actions viewed as defiance, and their authority undermined, particularly if their influence threatens civilian control or contradicts the government's chosen course.

Corollary of Maxim 4: A warrior who clashes with their politicians' objectives becomes an enemy of the state.

The true challenge, therefore, lies not merely in acquiring power but in wielding it with wisdom and foresight. Politicians and commanders alike must navigate the shifting political landscape and forge alliances that prioritize long-term stability over short-term gains. They must recognize the limits of military power and accept that true security can only be achieved through cooperation, mutual understanding, and restraint. Ultimately, the success of any political strategy depends on ensuring that the quest for security does not sow the seed of future conflict. Only then can the promise of a lasting peace be realized, not through the absence of power, but through its disciplined and judicious use.

CHAPTER 22

CASE STUDY

"So, this is how liberty dies—with thunderous applause."

— PADMÉ AMIDALA, SENATOR (LUCAS 2005)

In the *Star Wars* universe, no strategist surpasses Sheev Palpatine, known variously as Senator, Supreme Chancellor, Emperor, and Darth Sidious. His unmatched mastery of intelligence, deception, patience, charisma, ambition, and ruthlessness, combined with a profound connection to the dark side of the Force, secures his legacy as a strategist without equal.

Palpatine plays the long game, focusing on strategies to achieve long-term ends. He puts his plan in place prior to the events of *Star Wars: The Phantom Menace*. Later, he manipulated powerful people and groups to achieve his objectives. While he appears to have successfully manipulated everyone, his masterpiece is his control of Padmé Amidala and Anakin Skywalker.

As the duly elected queen and subsequent Republic senator representing her home planet of Naboo, Padmé invests her trust in the political mechanisms of the Galactic Republic. Simultaneously, Palpatine clandestinely orchestrates the Trade Federation's actions, leading to a military blockade and eventual occupation of Naboo, his own place of origin. Despite the mounting tension, Padmé persists in her pursuit of peaceful avenues for resolving the crisis, firmly asserting, "I will not endorse a path that steers us towards war" (Lucas 1999).

Though Padmé is skeptical towards certain Senate members and the Republic's bureaucratic apparatus, her unswerving faith in the Republic's ideals and a diplomatic resolution compels her to journey to Coruscant, the Republic's capital planet, to "present her case before the Senate." There she implores, "Honorable representatives of the Republic, distinguished delegates, and Your Honor Supreme Chancellor Valorum, I come to you under the gravest of circumstances. The Naboo system has been invaded by force. Invaded…against all the laws of the Republic by the Droid Armies of the Trade—"(Lucas 1999).

A Trade Union representative interrupts, challenging her assertions and labeling them as unbelievable because of insufficient evidence. They call for a commission to be dispatched to Naboo to ascertain the truth. Here, Senator Palpatine deftly maneuvers the political machinery of the Senate, employing delaying tactics. Later, in a private discussion with a confederate, Palpatine notes how he manipulates the political system to achieve their objectives, stating, "I have the Senate bogged down in procedures. By the time this incident comes up for a vote, they will have no choice but to accept your control of the system" (Lucas 1999).

Padmé realizes her failure and declares that the "Republic no longer functions." She returns to Naboo to lead an attack against the Trade Federation. The nine-year-old Anakin Skywalker saves her efforts by taking control of an interstellar fighter, destroying the Trade Federation's mothership, and decapitating the invading droid army. (Lucas 1999).

Unperturbed, Padmé returns to Coruscant as a member of the Republic Senate. She continues to fight for a political body that does nothing to save her people and that she declares ineffective, realizing the alternatives may be worse. There, Palpatine maneuvers Padmé to call for a vote of no confidence against the incumbent Chancellor Valorum, paving the way for the collapse of the Senate and the rise of Palpatine as the Galactic Emperor.

Palpatine manipulates the Senate into granting him emergency powers under the pretext of safeguarding the Republic and preventing thousands of planetary systems from seceding to form the Confederation of Independent Systems, often referred to as the Separatists. Without the Senate's knowledge, Count Dooku, Palpatine's coconspirator and Sith apprentice, assumes the identity of a deceased Jedi and clandestinely clones an army of genetically engineered human troopers from the DNA of the bounty hunter Jango Fett.

When they are discovered, the Jedi Council assumes command of these clone troopers, who become the core of the Grand Army of the Republic, to fight the Separatists and to preserve the Galactic Republic. This conflict, known as the Clone Wars, finds the Grand Army of the Republic (GAR) fighting the Separatists and their droid army, ultimately ending in a stalemate.

The Clone Wars play a pivotal role in enabling Palpatine's ascent to power by simultaneously weakening his rivals' military capabilities and instigating a corrosive influence on Anakin Skywalker's thoughts and allegiances.

In Padmé's defense, Palpatine also skillfully maneuvers the entire Senate, Jedi Council, and his new apprentice, the now-teenager and Jedi apprentice Skywalker. He manipulates Anakin's fragile ego and discredits the Jedi council leadership, so that when the time is right, Palpatine appoints Anakin to the Jedi Council, incensing the Jedi. The Jedi fall for his trap and don't promote Anakin to Jedi Master, further shattering his fragile self-esteem. Jedi leaders then ask him to spy on the one man who he believes is mentoring him. Palpatine consoles the now-disgruntled Anakin and plants the seed for sedition.

After her clandestine marriage to the now-nineteen-year-old Anakin and their subsequent pregnancy, Padmé assumes a pivotal role in Palpatine's meteoric ascent to authority. Anakin's attachment to Padmé becomes the linchpin in Palpatine's intricate manipulation of his emotions and desires. Anakin's premonition of Padmé's demise during childbirth, coupled with his overwhelming fear of losing both Padmé and their unborn child, propels him on a relentless quest for power and control.

Palpatine exploits Anakin's vulnerability, magnifying his inner turmoil and discontentment, offering Anakin the prospect of saving Padmé, securing Anakin's unwavering loyalty, and steering him to succumb to the dark side. This transformation culminates in Anakin's rebirth as Darth Vader, the ruthless enforcer of the new regime.

The Jedi move on Palpatine when they discover his treachery. However, Palpatine convinces Anakin that the Jedi are attempting a coup, and after Anakin kills a Jedi Master, Palpatine convinces him to eliminate the remaining Jedi. Padmé dies in childbirth, and even in her death, Palpatine uses her to convince Anakin that it is the fault of the Jedi.

Palpatine issues Order 66, which compels the clone troopers to systematically annihilate their Jedi commanders. He then commands Anakin to exterminate the remaining Jedi adults and children, destroy their temple, and execute the Separatist leaders. Then he issues a shut-down command to the droid armies.

In one fell swoop, Palpatine orchestrates a ruthless and calculated maneuver that eradicates his political and military adversaries and transforms the Galactic Republic into an Empire firmly ensconced under his control. The clone troopers eventually become the core of the Imperial Stormtrooper Corps under the Galactic Empire.

The Trade Federation went to war over economic interests, specifi-cally in response to Naboo's taxation of natural plasma exports, a vital resource used in advanced weapons. However, Senator Palpatine manipulated the crisis. He provoked the invasion to create a powerful Separatist movement and manipulated the Senate to execute a no-confidence vote against the incumbent Supreme Chancellor Valorum. In the power vacuum that followed, he positioned himself to be declared Supreme Chancellor. Once in office, he proclaimed martial law and suspended liberties, taking control of the Republic and trans-forming it into an Empire under his absolute control. While the Trade Federation fought for economic gains, and the Naboo forces and Jedi fought for freedom, Palpatine politically outmaneuvers everyone, using war as a tool to consolidate power and dismantle the Republic from within.

PART FIVE

ENDGAME

"It is easier to start wars than to end them."

— JEAN-LUC PICARD, CAPTAIN, USS ENTERPRISE
(NCC-1701-D) (SCHEERER 1992)

CHAPTER 23

PARLEY

"We're human beings with the blood of a million savage years on our hands, but we can stop it! We can admit that we're killers, but we're not going to kill today. That's all it takes…knowing that we're not going to kill today."

— JAMES T. KIRK, CAPTAIN, USS ENTERPRISE (NCC-1701) (PEVNEY ET AL. 1967)

A conflict ends when opposing parties mutually agree on terms to restore peace, or when one side unilaterally imposes its will on the other. When confronted with overwhelming odds, a pragmatic opponent may choose to surrender rather than risk a futile last stand, recognizing that continued resistance could lead to catastrophic loss and cripple any hope of recovery. In such circumstances, a society must weigh the value of resistance out of principle against the possibility of negotiating terms that preserve its people, history, and culture, even under occupation.

Even when one party believes it can eventually dominate the other, a negotiated peace can avoid the immense costs and consequences of a *decisive victory* and produce a more agreeable and therefore stable post-conflict environment. By prolonging lives, resources, and infrastructure that may otherwise be lost in prolonged fighting, negotiation offers a path towards recovery rather than devastation. It also enables both sides to communicate their needs and grievances, creating the possibility of mutually acceptable and enduring solutions.

Leaders may choose to negotiate a settlement or withdraw once they recognize that continuing the war no longer aligns with political realities. Negotiated conclusions are often more desirable than the outright destruction of an opponent, particularly when continued violence favors neither side's interests.

Maxim 56: Peace endures when it offers more than conflict.

Dialog and reconciliation occur when all parties believe they have more to gain through peaceful means than by continuing hostilities.

Beyond outright victory, there are two principal paths to ending a war. One is the negotiation of a settlement among all parties, which becomes viable when they conclude that a peaceful agreement offers a more favorable outcome than continued fighting, particularly as human, economic, and political costs rise to unsustainable levels.

Alternatively, a conflict may end without a decisive victor, with opponents going their separate ways without resolution. One side may withdraw to address more pressing challenges elsewhere, or public unrest or a collapsing economy may erode domestic support, compelling leaders to seek an end to hostilities. In such cases, the war effectively concludes even in the absence of a formal peace agreement. In this case, the parties may formally or informally agree to a ceasefire.

NEGOTIATIONS

To ensure a peaceful post-war environment, both parties should approach negotiations with an open mind and willingness to discuss, negotiate, and compromise. The outcome of a negotiation must be mutually agreed-upon terms that both parties will honor, which might involve making concessions such as territorial adjustments, *reparations*, or disarmament. Only by adopting a diplomatic approach can the parties expect to create a stable and peaceful post-war environment.

Establishing a *ceasefire* is often the first step in negotiating an end to hostilities. Halting violence allows both sides to implement additional terms of agreement, such as withdrawing troops or transferring resources. During this time, the treatment and repatriation of prisoners are often addressed, including their medical care, food, shelter, and a process for their return. Other forms of negotiations include:

- Peace treaty: a formal and enduring agreement intended to end hostilities and establish a stable postwar order.
- Truce: an informal pause in hostilities, usually of limited duration, that rarely reflects a willingness to resolve the broader conflict.
- Cessation of hostilities: a temporary agreement to halt fighting, without addressing the underlying political issues.
- Armistice: a formal and enduring agreement to cease all military operations, signifying the end of active warfare but not necessarily the establishment of a lasting peace.

At this early stage, the nature of war shifts from destruction to post-conflict governance. Yet the end of a war remains a perilous period. After a prolonged or brutal struggle, both sides may fear betrayal, increasing the risk of a double-cross. Leaders must be cautious when negotiating with an adversary that is not on the brink of defeat, collapse, or starvation, as such parties may exploit the negotiations to regain strength or shift the balance of power in their favor.

Negotiation is only necessary when an opponent still has the means to resist, since power defines possibilities, as noted by Maxim 1. Conceding irrecoverable assets, such as land, military capability, or technology, can leave a society dangerously exposed. These assets often provide the leverage that compels an opponent to negotiate or comply. Relinquishing them removes that leverage and increases vulnerability. Once an adversary's power is broken, the incentive to negotiate or to honor agreements quickly vanishes (Seabury and Codevilla 1990).

**Corollary of Maxim 1: Never trade power for promises.
Only the presence of power compels others to negotiate and to honor commitments.**

Even after a war ends or transitions into a prolonged ceasefire, unresolved tensions often persist. During this lull, either side may reposition, reinforce, or withdraw, treating the temporary peace as a strategic pause rather than a genuine resolution. Such interludes can be ruses, setting the stage for a renewed offensive aimed to "finish this once and for all" in a decisive battle.

**Converse of Maxim 11: Never underestimate a defeat; your opponent may return stronger.
A defeat must not appear temporary, or the opponent will endure the setback and strike when it is prepared.**

The successful transition from war to a stable peace requires careful planning, cooperation, and unwavering commitment from all stakeholders. However, societies often discover that the end of a war is not the end of a conflict; it is just the beginning of the process of rebuilding and healing on both sides. Politicians must work toward creating a safe and stable environment to prevent future outbreaks of violence or instability, and to prevent the formation of rebel or fringe groups that may threaten the peace.

Rebuilding often begins with repairing or replacing society's critical components, such as utilities, hospitals, and government facilities. These essential systems must be repaired or rebuilt to reestablish normalcy and enable the recovery of shelter, food, water, and clothing. Leaders must also confront the economic fallout, including unemployment, destroyed industries, and lost income, as they shift from a wartime to a peacetime economy. This includes demobilizing war industries, retraining former war workers, and jumpstarting economic growth to improve living conditions.

At the same time, leaders must restore the broader forms of societal power fractured by conflict. Political institutions must be rebuilt to reestablish legitimacy and maintain order. Law enforcement, courts, and administrative bodies need to regain both operational capacity and public trust. Education, media, and cultural institutions must also be revitalized to promote stability, a shared sense of identity, and long-term resilience. Without these political, legal, economic, and social sources of power, a society remains fragile and unstable, even after the violence ends.

CONCLUSION

Long after a war has formally ended, tensions may continue to simmer between the former adversaries. A mutually agreeable solution not only lessens the cost of war but also lays the foundation for enduring peace, reconciliation, and healing. Preventing future conflict demands that the root causes of the war, whether political, social, or economic, be addressed through meaningful reform and dialogue. Only by resolving these underlying divisions can former enemies transform hostility into stability and ensure that they do not find themselves at each other's throats in the future.

CHAPTER 24

VICTORS AND THE VANQUISHED

"War tends to distort our point of view. If we sacrifice our code, even for victory, we may lose that which is important for our honor."

— OBI-WAN KENOBI, JEDI MASTER (LEE, LUCAS, AND MICHNOVETZ 2011)

A group achieves a victory by overwhelming their opponent's physical capacity to fight or crushing their will to resist. The first approach involves disabling or destroying an opponent's war machine or support infrastructures. The second approach involves creating a sense in their opponent that their defeat is inevitable, encouraging them to surrender since further resistance and bloodshed is pointless.

Maxim 57: A victory requires eliminating physical or psychological resistance.
An opponent may capitulate when no other viable options remain.

A victor may demand an *unconditional surrender*, dictating terms that include disarmament and troop withdrawal. More importantly, the victor sets the terms of peace, shaping the future of the defeated society, including territorial concessions, reparations, disarmament, or even fundamental changes to political, economic, and social systems. In essence, victory allows the victor to impose their will unilaterally on the defeated.

For a reasonable victor, surrender can provide a foundation for peace and reconstruction, offering the defeated society a chance to survive under new conditions. But in the hands of a ruthless conqueror, surrender becomes a mechanism for domination and cultural erasure. In such cases, resistance, however costly, may seem the only means to preserve dignity or prevent atrocities. The choice becomes a perilous crossroads, where surrender could lead to hope or to annihilation.

Yet passion often overrides pragmatism. Civilizations facing extermination, the obliteration of their heritage, or the desecration of sacred lands may see surrender as unacceptable. In these cases, resistance may continue out of principle, vengeance, or desperation. One final, violent stand may be seen not only as a refusal to submit, but to be remembered, to delay the enemy, or to ignite future resistance.

Fanatics such as Khan Noonien Singh from *Star Trek* and Paul Atreides from *Dune* are driven by unwavering conviction. Those fueled by religious fervor, a belief in destiny, or an unrelenting desire for vengeance might prioritize inflicting maximum damage or achieving a glorious end, even in the face of certain defeat. For such individuals, the notion of surrender holds no appeal. Instead, they may embrace a last stand that cripples their enemy even as it ensures their own destruction.

Synthetic intelligences like the Berserkers from the *Berserker* series and Skynet from *Terminator* lack the capacity for empathy or the concept of surrender. Their programming might prioritize relentless conquest or self-preservation above all else, making compromise or negotiation impossible. These entities may view all beings as obstacles to be eliminated, leaving only complete destruction as a solution.

Alien creatures like the Arachnids from *Starship Troopers* and the Xenomorphs from *Alien* may possess indomitable spirits and show no openness to negotiation or to the concepts of retreat or defeat. Their behavior suggests instinctive rather than strategic decision-making, which renders traditional diplomatic or psychological approaches ineffective. Their actions may be driven solely by the need to consume resources or fulfilling a biological imperative, resulting in a relentless, fight-to-the-death mentality.

Ruthless dictators, like Palpatine from *Star Wars* and Big Brother from *Nineteen Eighty-Four,* may be immune to their people's suffering. Even if defeat could minimize their citizens' suffering, the prospect of losing control would drive them to fight relentlessly.

Finally, a warrior race like the Klingons from *Star Trek*, and both the Fremen and Sardaukar of *Dune* value strength and honor above all else. They view surrender as a sign of cowardice and believe a glorious death in battle is preferable to defeat and enslavement. Traditional warfare tactics and negotiations focused on preserving life may be ineffective against such opponents. A victory against one of these foes may require the destruction of their means and of their society.

A victor may decide that it will not negotiate nor accept its opponent's surrender. It may decide that the destruction of the other's means to resist, or their entire society, is the only acceptable end. This is especially true if they began a war to punish their opponent for perceived past injustices, to obtain property or resources through conquest, or to act out passions fueled by racial, religious, or philosophical conflicts. The victor, convinced of the righteousness of their cause, might believe the complete eradication of their opponent is the only acceptable solution.

The ends envisioned at the beginning of a war, whether protection, passion, property, punishment, or power, paint a picture of what victory might look like. Yet, war can obscure these expectations. Victors may find the post-war landscape far more complex and the true costs of achieving their ends far steeper than they had expected.

THE VICTORS

"No dictator, no invader, can hold an imprisoned population by force of arms forever. There is no greater power in the universe than the need for freedom. Against that power, governments, tyrants, and armies cannot stand."

— G'KAR, SPEECH TO THE BABYLON 5 ADVISORY
COUNCIL (STRACZYNSKI 1995B)

A victor must determine how to deal with a defeated foe. Should it occupy the conquered territory or withdraw? Offer assistance or impose punishment? Reinstate the former leadership, establish a new regime, or abandon the society altogether? The victor may also demand reparations, framed as justice for aggression or compensation for losses, though such demands are often little more than extortion. These decisions shape both the legacy of the war and the stability of the post-conflict order.

If a victor's objective is conquest, then they must develop a postwar plan for dealing with the population. A victor is more likely to preserve a conquered people if it can be incorporated into the victor's society and they contribute value. One approach is to establish a proxy or puppet government, installing native collaborators or loyal subordinates to govern the population on the victor's behalf. However, to maintain control, the conqueror may need to infiltrate the new government with overseers, spies, and collaborators. This method allows the victor to extract resources and maintain order with minimal effort, while permitting the conquered population to retain elements of their own culture.

Alternatively, the victor may *annex* a territory, thereby assuming ownership of the land and its people. In this case, the conquered become part of the victor's society, often creating a culture of subservience while expanding the victor's domain. However, a population may resent governors imposed by distant rulers.

An extreme form of annexation is assimilation, in which the defeated society is completely absorbed into the victor's culture and identity. In the *Star Trek* universe, the Borg exemplify this concept by assimilating life forms and technologies into their collective consciousness, erasing individuality in favor of unity and efficiency. Their actions raise profound ethical and moral questions, not only about the nature of the Borg themselves, but also about how to respond to their horrific strategy.

When the Enterprise crew encounters a lone survivor of a crashed Borg scout vessel, many want to execute the Borg before it can call its immensely powerful mothership. Instead, the medical officer insists they transport the lone Borg aboard the *Enterprise*.

As the Borg survivor recovers, Captain Picard announces a plan to infect the Borg-human hybrid with an "invasive programming sequence," or computer virus, and then release the survivor, in the belief that it would infect the entire collective, shut down their collective mind, and eliminate the Borg menace (Echevarria 1992).

Many of the crew feel this action is immoral biological warfare, and a deliberate attempt to eliminate the Borg species. They argue that intentionally infecting the Borg with a virus is wrong, even if it could eliminate them as a threat. However, Maxim 41 warns that horrific enemies often compel horrific choices. Given Picard's traumatic experience of being assimilated and subsequently rescued from the collective, it's understandable that he views the Borg as an existential threat to his crew and the entire galaxy.

Picard does not regard the Borg as a distinct species, but a relentless, aggressive force that endangers the freedom and individuality of countless beings across the galaxy. Consequently, he perceives his actions not as xenocide, but as administering a cure for a pandemic.

In the future, warriors may grapple with profound ethical dilemmas about the acceptable measures they can employ when dealing with individuals from diverse societies. These quandaries become especially pronounced when they face adversaries that intend to either annihilate or subjugate their race. What limits should a leader impose against an opponent bent on their society's conquest or eradication? What should a leader do if an opponent mobilizes its entire species for war? Does every member of that race automatically assume the role of a combatant, thus rendering them viable military targets?

How should political and military leaders address the potential eradication of an entire race?

In the end, Picard decides that "To use [the Borg survivor] in this manner would be no better than the enemy we seek to destroy" (Echevarria 1992).

This decision reveals a central paradox of war: to justify violence, warriors must often dehumanize their enemies, casting them as monstrous threats deserving annihilation. Yet to restore peace and achieve reconciliation, they must later rehumanize those same foes. This duality creates a deep psychological conflict, often leaving behind resentment and generational hatred that can destabilize postwar societies and complicate the governance of conquered worlds.

After a victor captures a society, the fate of the defeated military and paramilitary forces is a pivotal consideration. These individuals, trained and equipped to safeguard their society and uphold its political system, pose a risk of future resistance. To mitigate this, the conqueror may disarm or disband the vanquished security forces. However, the repercussions of such actions can be profound, especially if the society has already endured substantial damage during the conflict. Dismantling the institutions that enforce law and order may trigger societal collapse and usher in a dangerous period of anarchy. This power vacuum creates fertile ground for increased crime and societal chaos. In such a scenario, opportunists may exploit the instability to seize control, potentially ushering in a new, brutal regime.

A victor may conclude that establishing a robust military presence is essential to maintain security and remind the vanquished of their place. Yet such visible displays of dominance through bases, patrols, and checkpoints, though intended to discourage conflict, can produce the opposite effect. Rather than providing stability, these symbols of oppression become constant reminders of defeat, loss, and humiliation. Over time, they can foster resentment and inspire renewed resistance and rebellion.

Insurgents may target these conspicuous reminders, aiming to undermine the occupier's authority. The population may also attempt to corrupt or influence the occupation forces by appealing to their compassion, or through bribes, threats, or propaganda.

In **Avatar**, Jake Sully *goes native* on Pandora by gradually adopting the customs, values, and identity of the alien Na'vi people, abandoning his original loyalty to the human military-industrial mission. As he trains with Neytiri and lives among the Omaticaya, he forms emotional bonds, takes part in spiritual rituals, learns their language, and ultimately embraces their worldview. He rejects the human plan to exploit Pandora's resources and leads the Na'vi in resisting the invasion. To the Na'vi, he becomes a trusted ally and a great friend. To many of the humans, he is a horrific traitor to humanity, showing that loyalty and identity follow Maxim 14, depending on one's point of view.

A victor may continue to punish the defeated to remind them of their humiliation, to "keep them in their place," and to ensure the victor's own protection and prosperity. This is the case in **The Hunger Games**, where the powerful Capitol controls twelve districts. As punishment for a previous rebellion, the Capitol uses an annual lottery to select one male and one female child from each of the districts and forces them to fight to the death in a televised battle royale. The Capitol broadcasts the Hunger Games nationwide and forces the districts to watch as their children fight to survive in a brutal, deadly arena. The Hunger Games reminds the districts of the Capitol's power and control over them, using fear to deter future rebellions. However, it does the opposite.

A long-term solution might involve establishing a religion or other institution to control and guide a populace. Asimov's **Foundation** series involves a religion of science that unilaterally controls the maintenance and operations of all science and technology, including nuclear power. When human civilization collapses, the religious leaders control electrical power generation and nuclear-powered spacecraft. This authority enables them to control political and military leaders. When political leaders become dependent on religious leaders, they must ensure those religious leaders' loyalty or risk becoming puppets themselves.

Conquerors may choose to remove a population to secure future peace, facilitate pillaging or colonization, or create a depopulated buffer zone. A conqueror is more likely to pursue removal when the war was not fought to liberate an oppressed society, when the two societies hold incompatible interests or ideologies, or when the victor has ample resources and no need for additional labor. If the conquered are seen as resources themselves, such as food or fertilizer, their fate is often sealed.

One approach to removal is exile. A victor may banish a people by expelling them without assigning a destination, turning them into nomads, or forcing them to seek a new homeland. Alternatively, they may displace them by relocating the population to a designated area. Both options can be difficult to manage, as these groups may become unpredictable and hostile. They may impose logistical burdens on the victor, who may feel compelled to provide food, shelter, and transportation.

Another option is to impose internal exile, which functions like a bypass strategy. Instead of relocating the population, the conqueror isolates them within their own territory by dismantling transport infrastructures or using surveillance, outposts, and patrols to restrict movement. This approach effectively transforms their homeland into a prison without walls.

Once uprooted, a population may find itself progressively displaced to less favorable lands, leading to its decline or dependence on external aid. Those displaced often hold fast to their claim to their ancestral homeland, a conviction that fuels lasting resentment, unrest, and enduring challenges to peace.

In *Dune*, the Fremen of Arrakis endure centuries of exploitation and displacement under successive regimes that seek control of the planet's valuable spice. Despite their marginalization, the Fremen retain their identity and deep attachment to their homeland, eventually rising to overthrow their oppressors and reclaim Arrakis.

In the original *Star Trek* universe, Captain James T. Kirk defeats the genetically engineered superhuman augment, Khan Noonien Singh, and exiles him, along with his followers, to the uncolonized world of Ceti Alpha V (Wilber, Coon, and Wilber 1967). In *Star Trek II: The Wrath of Khan*, we learn that at the time of their exile, the planet is only marginally habitable, but six months later, the neighboring planet Ceti Alpha VI explodes, shifting the orbit of Ceti Alpha V and transforming it into a barren desert world. This catastrophe brings prolonged hardship, starvation, and death, including the loss of many of Khan's followers and the eventual death of his wife, Marla McGivers, as a result of the deadly native Ceti eels. After fifteen years of suffering and bitterness, Khan seizes his opportunity for vengeance when he again encounters the Federation (Meyer 1982).

Compassion is not universally embraced, and its absence can produce warriors conditioned to eradicate their enemies without hesitation. As early as 1898, H. G. Wells explored the evils of colonialism against "inferior races" and the logic of extermination in *The War of the Worlds* (Wells 1898), in which Martian invaders employ ruthless tactics against humanity. The Martians show no interest in negotiation, assimilation, or mercy. They possess overwhelming technological superiority, including heat-rays that incinerate targets instantly and black smoke, a poisonous gas that kills indiscriminately and spreads terror across the countryside. Their strategy mirrors the worst aspects of imperial conquest, prioritizing domination and resource extraction over any consideration of coexistence. The Martians destroy infrastructure, decimate populations, and dismantle organized resistance with clinical efficiency. This is the pursuit of victory without empathy, rooted in annihilation rather than subjugation, where the enemy is not merely defeated but erased.

Some races may find it difficult to be instruments of genocide. Others may see genocide as the only rational means of handling a threat. Consider a leader who orders a siege and demands that their opponents surrender. If the opponents refuse, then the attacker may face a long and costly campaign to defeat their defenses and population. To prevent this from occurring and recurring, attackers may adopt a "no quarter policy," in which they eliminate some or all of the defenders and inhabitants for resisting their attack. The leader may rationalize this policy, believing it will avoid future bloodshed by motivating others to surrender. They may see themselves as placing the lives of their troops before the lives of their opponents, while others may view their strategy as morally and ethically abhorrent.

Likewise, when every member of a society is a combatant, a leader may see no way to defeat this opponent without forcing an atrocity. In *Ender's Game*, Andrew "Ender" Wiggin unknowingly eliminates the insectoid alien race known as "Buggers," and is heralded as a hero to humanity. However, he struggles for the rest of his life to atone for his extermination of the species.

While annihilating a defeated society may remove the burden of occupation, oversight, and potential rebellion, the consequences extend far beyond the immediate destruction of life and culture. Those who commit such acts may face severe legal, political, and social repercussions, as well as near-universal condemnation. Genocide often drives survivors and their allies to fight with desperate resolve, believing they have nothing left to lose. In many cases, it ends disastrously for those who initiate such violence. Though genocidal or xenocidal policies may seem to eliminate a threat in the short term, they may create lasting hatred, instability, and vengeance that endures for generations.

In *Battlestar Galactica*, the Cylons' surprise attack results in the mass extermination of humanity's population centers. Later, when some Cylons seek to end hostilities, the surviving humans reject their overtures, unable to trust those responsible for the initial assault, the massacre of billions of civilians, the destruction of cities, and the repeated use of saboteurs and assassins against political and military leaders Such atrocities carry lasting consequences, demonstrating that the use of horrific means inevitably shapes the prospects for reconciliation and peace.

CONCLUSION

"Tell a conquered man he has a new master, and he'll shrug. Tell him his new master wants a fifth of his annual income, and he'll go and find his pitchfork."

— HORUS LUPERCAL, PRIMARCH OF THE LUNA
WOLVES (ABNETT 2006)

A leader may discover that winning a war and conquering worlds is the simple part. The genuine challenge begins after the fighting ends: consolidating power, maintaining control, and governing a wounded and resentful population. Victory on the battlefield does not guarantee lasting authority or prosperity. A newly conquered society may face economic collapse, infrastructure failure, civil unrest, and deep cultural resistance, revealing that the struggle for peace can be as demanding as the war itself.

The occupying force must restore essential services, establish a legitimate governing structure, and address the grievances of the local population, all while suppressing potential insurgencies and preventing rival factions from filling the power vacuum. This can prevent the rise of new threats and promote lasting peace and security, but it often demands significant resources, time, and political will. Or, a victor may choose not to rebuild, citing limited resources, domestic opposition, or the belief that reconstruction is not their responsibility. While this approach avoids the costs of occupation and counterinsurgency, it comes with serious long-term risks. Demands for restitution, taxes, or tribute become meaningless if the defeated lack the infrastructure to comply. A society left in ruins may collapse into lawlessness, fueling unrest, insurgency, or criminal networks. In time, these conditions may generate new conflicts more dangerous than the original war. Without a clear and sustainable plan for postwar governance, even the most decisive military triumph can unravel into chaos and prolonged instability.

Long-term peace may only be possible if the victor addresses the underlying root causes of a war, if the people view the peace as fair and just, and if the terms of peace promote long-term stability and prosperity on both sides. Failure to address these underlying issues may lead the victor to perceive the defeat of the enemy's military as the ultimate triumph, while the defeated party might perceive this confrontation merely as the opening salvo in an extended conflict.

Even after having their security forces neutered and enduring a foreign occupation, some leaders, societies, or species may never capitulate. Bitterness and dissatisfaction may turn a victory into a bloody guerrilla war. And every time an insurgency movement destroys property or harms people, it generates additional anger and costs that make future reconciliations more difficult.

This dynamic is exemplified in the Rebels of *Star Wars*. The various Rebel factions are united by a deep-seated resentment toward Palpatine's manipulation of galactic leaders and his systematic dismantling of democratic institutions. He demonstrates his willingness to use ruthless authoritarian ways when he orchestrates the elimination of the Jedi, devastates the Separatists, and proclaims himself Emperor. Palpatine employs martial law and fear, even resorting to the mass killing of billions of his own citizens.

His despotic and totalitarian rule becomes the catalyst for the Rebels' determination to resist and overthrow imperial control. The Rebels are inspired to fight against the oppressive regime, driven by a profound hatred and the need to restore justice, democracy, and fundamental rights that are suppressed under Palpatine's tyrannical rule. Star Wars illustrates how unaddressed grievances and oppressive rule can perpetuate a cycle of conflict, where military victory alone does not ensure a lasting resolution.

CHAPTER 25

THE ASHES OF VICTORY

"The war does not end when people put down their guns. It ends when they reconcile. Until then, the war has only paused."

— CHRISJEN AVASARALA, EARTH'S SECRETARY-GENERAL OF THE UNITED NATIONS (EISNER 2019)

At the conclusion of hostilities, it is common for participants to realize that their original objectives and ends remain unfulfilled. The vanquished may be compelled to accept military and political consequences they originally found intolerable, compounded by death, suffering, territorial loss, economic collapse, reduced sovereignty, and the enduring humiliation of defeat.

Likewise, victors rarely experience the rewards they imagined at the outset of hostilities. In hindsight, they often find that the misery, death, destruction, disease, starvation, and immense cost of war bear little resemblance to the glory or advantage they once sought.

Nor is the end of hostilities the end of a conflict. The conditions that follow may fail to establish lasting peace or may even hasten future conflicts. The suffering endured during combat and occupation, combined with the absence of genuine reconciliation, can leave behind deep resentment and unresolved grievances. In such cases, war becomes the way to an end, though rarely the one envisioned at the outset.

As the World Coordinator notes in Isaac Asimov's, *I, Robot*:

Every period of human development has had its own particular type of human conflict—its own variety of problem that, apparently, could be settled only by force. And each time, frustratingly enough, force never really settled the problem. Instead, it persisted through a series of conflicts, then vanished of itself ... "not with a bang, but a whimper," as the economic and social environment changed. And then, new problems, and a new series of wars. (Asimov 1950)

Leaders often discover in hindsight that war creates as many, if not more, difficulties than it resolves. A war will wreak havoc on the battlefield and the people that lived and fought there, resulting in deaths and crippling injuries, destruction of military resources and civilian infrastructures, loss of state treasures, and missed economic opportunities. This may lead to costly demands for reparations for the destruction on both sides.

However, the costs of war are not confined to the battlespace. In both victory and defeat, war exposes the tension between steadfast ideals and harsh realities. In hindsight, a population may conclude that the ends were not worth the price paid. Grieving parents may ask, "Did my child's sacrifice mean nothing?" Political opponents may argue that their leaders should have known better, prepared better, led better, and protected the people more effectively. War compels societies to reevaluate deeply held political, economic, cultural, and religious convictions. In doing so, it often weakens the very foundations of power and stability, setting the stage for profound societal transformation. Whether a society emerges stronger or collapses entirely depends on how it navigates this test of faith.

Likewise, a victor may dismantle the institutions that once sustained the defeated regime, creating a power vacuum that paves the way for civil unrest. Without a functioning authority, opportunists, whether politicians, warlords, or criminal organizations, may seize control under the guise of restoring order or pursuing power. In such conditions, Maxim 1 reasserts itself, as the strong exploit the weak and divide the remnants of a broken society to serve their own ambitions.

Maxim 58: Power abhors a vacuum.
Power will always seek the weak.

Violence often leaves a deep, smoldering hatred in the hearts of those directly involved. A society may deliberately demonize its opponents to justify starting or engaging in a conflict, to spur patriotism and motivate recruitment, or to justify domestic actions such as confiscating property, raising taxes, or restricting rights. As conflicts drag on, one side's acts of desperation or expedience can become the new norm, creating a spiral of escalation in which former limits on violence collapse. Atrocities such as surprise attacks or the killing of prisoners may expedite the end of a war but plant the seeds of resentment that endure for generations. Such hatred does not vanish with the signing of a peace treaty.

In the *Lost Fleet* universe, Captain John "Black Jack" Geary awakens after nearly a century of cryogenic suspension. He discovers that during his slumber, the Alliance and Syndicate Worlds, two interstellar empires locked in a brutal war, have disregarded many of humanity's rules of warfare for the sake of expediency and vengeance.

Alliance leaders have learned to leave enemy troops to die on crippled spacecraft and to destroy ships that cannot defend themselves, including unarmed survival pods. Warships drop kinetic rounds on planets and target cities, and the noncombatants sheltered there. These atrocities horrify Geary, and he convinces his subordinate leaders that the evil of their actions escalates violence without honor and fuels the hatreds that sustain the war (Campbell 2007).

After the threat of conquest or destruction has passed, survivors may reexamine the violence they endured and question the methods used, along with those who ordered and performed the acts. For those who witnessed atrocities or lost loved ones, forgiveness or forgetfulness rarely comes easy. Meanwhile, those who ordered or took part in these acts may have been convinced of the righteousness of their cause and the evil of their opponents. Clinging to these beliefs allows them to justify their actions, making it difficult, if not impossible, to view former enemies as victims or as individuals worthy of trust or compassion.

When a civilization unleashes bigotry, ethnocentrism, or xenocentrism to demonize its enemies and rationalize violence, those beliefs do not simply vanish when the war ends. This is especially true in protracted conflicts where generations have fought, killed, and died. Demonizing the opposition provides a convenient moral justification for wartime atrocities, but such tactics often backfire. Victims, shaped and fueled by the same narratives once used against them, are more likely to seek vengeance after the war (Grossman 1995). The result may be a lasting barrier to peace and reconciliation.

Others may also examine a conflict in hindsight and come to view specific incidents, or an entire conflict, as criminal acts. In the aftermath, leaders may pursue justice and accountability by holding warriors responsible for war crimes. The victors may choose to punish those who ordered, committed, or enabled atrocities as a visible effort to seek atonement and reinforce moral authority. Alternatively, a society may choose to overlook these crimes if prosecution risks reigniting conflict or destabilizing a fragile peace.

Then again, shame, disgust, and compassion may not be universal emotions.

Corollary of Maxim 1: Winners determine what constitutes an atrocity or a war crime.

WAR'S ENDURING COSTS

"It is a great irony that the nature of war always reveals the true nature of those who fight."

— CORTANA, UNSC SYNTHETIC INTELLIGENCE
(FUTAMURA 2010)

The most enduring consequence of war may be the vast diversion of resources from the sustenance and advancement of society. These costs extend far beyond the immediate expenditures on weapons and battle-field destruction. Instead of contributing their skills to constructive pursuits, soldiers, scientists, engineers, and innovators support the war effort, limiting advancements in non-military fields. War also necessitates the reallocation of resources away from vital areas such as education, healthcare, and social development, hindering a society's capacity to invest in its people and its future. Over time, this diversion can weaken the economy, reduce social mobility, and diminish innovation. While the immediate costs of war are undeniable, it is the lost potential and deferred progress that prove to be its most enduring consequences.

A conflict may also create unintended consequences on both sides. Consider the long-term threat of orbital debris from space combat. Unlike projectiles on Earth, those launched in space can remain on a potentially catastrophic trajectory for centuries, until drawn by the gravity of another body or striking another object. Missed shots, mines, and damaged spacecraft can create vast debris fields orbiting planets, posing a danger to functioning satellites and spacecraft navigating these zones. The perpetual tug of gravity may eventually drag damaged and destroyed satellites, spacecraft, weapons, and debris to rain down without warning onto a populace. Space warfare has the potential to create a self-perpetuating hazard, jeopardizing not only immediate combatants but also future generations and the infrastructures they rely on.

Explosives and orbital bombardments may throw chemicals, radiation, and other debris into a planet's atmosphere, water, and soil. A strike on a nuclear power plant will spread radioactive debris around the impact area and into the atmosphere. A strike near a chemical refinery or storage facility may create a toxic plume that spreads with the wind. Over time, this airborne debris and radioactive fallout will rain down and contaminate crops, cities, and the populace. A military victory might still result in devastating losses as the populace is slowly poisoned, starved, or left with an environmental disaster.

In *Star Wars*, the destruction of the second orbiting Death Star battle station and its subsequent impact with the moon planet of Kef Bir causes an environmental catastrophe. Later, when the Empire is in its final death throes, Imperial and New Republic forces conduct a last-ditch battle on and around the planet Jakku, littering the planet with wrecks, debris, and toxic waste (Abrams, Kasdan, and Arndt 2015; Wendig 2017). Even as the Empire crumbled, the seeds of future conflict were sown amidst the wreckage of Jakku, a grim reminder that war often begets war.

Taken to its extreme is the nightmarish future of **Warhammer 40,000**, where humanity is locked in perpetual war. Millennia of conflict have shaped societies into brutal war machines, with the vast Imperium of Man prioritizing military dominance and religious zealotry over progress. Stagnant and on the verge of collapse, the Imperium consumes entire planets for industrial production to fuel its war machine, leaving billions to live in fear and squalor. The constant threat of alien invasion forces humanity to prioritize endless war at the expense of any hope for a better future (Abnett 2005).

THE INVISIBLE WOUNDS

"In time of war, we would have never left a man behind."

"Maybe that's why we lost."

— ZOË ALLEYNE WASHBURNE AND MALCOLM
REYNOLDS, SERENITY (WHEDON 2005)

Maintaining a military force is expensive, especially when the threat does not justify the cost. As a result, at the end of a conflict, a society may choose to sell, store (e.g., mothball), or discard military equipment, while demobilizing, disarming, and reintegrating its combatants into civilian life.

Societies train warriors to use violence, often reshaping their views on societal norms. However, witnessing and perpetuating violence can have a profound impact on an individual's psyche. In taking the life of an enemy, a soldier may kill a part of themselves, resulting in psychiatric trauma (Grossman 1995). The honors that a society awards its soldiers does not serve as an ego boost, but as a societal acknowledgement that their actions, no matter how violent, were morally justified.

However, veterans may find themselves treated dishonorably. In *Firefly*, Captain Reynolds and First Mate Washburne are veteran "Browncoats" of the Unification War. This conflict pitted the Alliance, a powerful central government, against the Independents, who opposed the Alliance's control and sought to maintain their independence. Having found themselves on the losing side, the crew of the *Serenity* must continually deal with the effects of the war. Citizens and veterans of the Alliance treat them as traitors, the Alliance government harasses them, and they are forced to deal with their own loss and psychological issues from the war.

Firefly portrays the struggles of veterans attempting to reintegrate into society after violence has become their means of survival. Abandoned, disillusioned, and questioning their sacrifices after being "drummed out" of service, they must face their trauma in isolation. For many, violence becomes an instinctive response to a world that no longer understands or accepts them.

To avoid this, a society may need to educate, retrain, and reintegrate returning warriors. Veterans may require pensions and medical assistance, which can place additional strains on society.

On the other hand, military service can elevate warfighters to prominence among the populace, giving them a platform to launch their own political careers, challenge existing political structures and leaders, shape public opinion, and threaten established politicians and parties. Veterans may advocate for peaceful political change or for the use of military force to coerce politicians and the populace.

Corollary of Maxim 1: Care for those who understand violence, or eventually they will take care of themselves.

Politicians who fear popular military leaders may take actions that create a self-fulfilling prophecy. In *The Lost Fleet* universe, Geary returns after defeating his society's long-standing enemy. Though loyal to his elected government, the populace hails him as a savior for ending a devastating century-long war with the Syndicates. In a democracy, such admiration grants him tremendous, if unwanted, political power. Alarmed by his growing influence, politicians send him on a perilous mission into uncharted space, giving him vague and contradictory orders amid the threat of unknown alien forces. In their desperation, they reignite the war not to defend the people, but to eliminate the threat posed by a powerful military leader and to preserve their fragile hold on power.

Alternatively, a civilian populace may despise veterans for their perceived violent and uncivilized behavior. This may lead to abandoning veterans, calls for the prosecution of military leaders, or investigations into illegal and unethical promotions or exemptions from service. Or there may be witch-hunts simply to make or break careers.

In *Old Man's War*, the Earth government adopts an unconventional approach to reintegrating returning warriors into society: they are not permitted to return to the world they left behind. Those who enlist in the interstellar conflict are told from the beginning that their ties to Earth and to their former lives will be permanently severed. In exchange for having their minds transferred into healthy, youthful bodies, the government relocates surviving veterans to distant outpost colonies. This places warriors on humanity's border and keeps these enhanced, battle-hardened veterans separated from mainstream society.

Mercenaries and privateers face additional challenges and opportunities. Once a threat has been eliminated, its sponsor may decide that its services are no longer necessary, leaving it unpaid and seeking alternative employment. As politicians decommission conventional military equipment and retire veterans, these mercenaries, privateers, and pirates may find themselves the most potent or sole remaining military force, creating opportunities for them to assert dominance and seize what they need or desire.

War can also create lucrative opportunities for those who provide military equipment, supplies, training, and transportation, as well as for those who control contracts to acquire these items and services. In the aftermath of a conflict, as the urgency of the situation subsides, investigations into war profiteering, price gouging, graft, and corruption may emerge. In the world of *Ender's Game*, some businesses prioritize maximizing corporate profits over addressing an imminent existential threat to Earth and humanity. When corporate leader Ukko

Jukes and his son capture alien technology and information offering the potential for immense wealth, they face a moral quandary: should corporate gain take precedence over the survival of their species?

CONCLUSION

"Wars not make one great."

— YODA, JEDI MASTER (LUCAS 1980)

"Violence is the last refuge of the incompetent."

— SALVOR HARDIN, MAYOR, TERMINUS (ASIMOV 1951)

War is a brutal catalyst for change. Yet even in victory, and with the change that comes with it, lasting peace is not guaranteed. The vanquished may be coerced to accept what they once resisted, including the loss of freedom, wealth, lives, and dignity. Such humiliation can deepen resentment and lay the groundwork for future violence.

If an invader's ends were to gain land, resources, or regime change, then a lengthy occupation may be necessary. This may involve removing unwanted individuals and establishing a new government. Likewise, a war that devastates cities, agriculture, industries, and infrastructures may leave a legacy of suffering, hatred, and desires for revenge on both sides, making true peace elusive.

It is convenient to attribute war solely to the failure of politicians to achieve their desired ends through peaceful means. However, as shown, wars also arise from enduring hatreds, narcissistic rulers, unchecked ambitions for wealth, power, and glory, and grievances left unresolved from previous conflicts.

War is also a harsh teacher. Victors, blinded by success, often cling to their proven tactics and weaponry, reluctant to change means and ways that once brought victory. The defeated are forced to confront their shortcomings and learn invaluable lessons. They may uncover weaknesses in their strategies, tactics, weapons, or leadership that drive them to adapt and prepare for the next conflict.

Even bystanders learn invaluable lessons from war. Just as a species that cannot adapt to change faces extinction, a society that cannot adjust to evolving threats risks being defeated. Observers may study a conflict to learn new means and ways, growing stronger without suffering losses. In some cases, these third parties may view a recently concluded conflict as a prime opportunity to move in and sweep away the weakened combatants.

Corollary of Maxim 25: Success encourages complacency. Failure inspires ingenuity.
A stagnant society will eventually be defeated.

War's final toll extends far beyond the immediate battlefield. The cascading disruptions it triggers create a web of missed financial and personal opportunities. From stifled innovation to a decline in social mobility, these ripples may weaken nations for decades. While the wounds of war may heal, the lost potential for progress leaves scars not only on those in the present but also on generations yet to come.

EPILOGUE

"Y'know, there will be a time when you will need to remember that no matter how bleak or unwinnable a situation, as long as you and your crew remain steadfast in your dedication, one to another, you are never ever without hope."

— JEAN-LUC PICARD, ADMIRAL, STARFLEET
(SCHEERER 1992)

This exploration of military strategy has extended into the boundless realm of science fiction, revealing enduring military concepts that transcend time and context. The importance of intelligence gathering and situational awareness, the need for clear and effective communication, and the value of building strong and resilient military organizations and alliances are just a few of these. These principles extend beyond the realm of science fiction. They are the threads that weave together the fabric of effective military strategy across all realities.

A successful strategy involves more than selecting the appropriate means and applying them effectively to achieve desired objectives and ends. It is woven into a society's cultural fabric, shaped by its past triumphs and setbacks, and guided by its collective dreams and aspirations.

Ultimately, this exploration demonstrates that war is a complex and evolving phenomenon that requires careful study and genuine understanding. Science fiction serves as a valuable lens through which to examine how war both influences and is influenced by the intricate interplay of technological and cultural forces.

Consider *The Expanse*, a prime example of how science fiction portrays the multifaceted nature of war. Violent conflict is just one thread in a complex tapestry woven from the intricate interplay of politics and culture within a colonized solar system. Earth, Mars, and the Belt find themselves locked in a precarious dance of conflict and cooperation. Each faction pursues its own self-interest, leading to a state of perpetual tension punctuated by outbreaks of violence.

Science fiction also highlights how the success of future leaders hinges on their adaptability. Technology advances at an unprecedented pace, geopolitical dynamics shift with astounding rapidity, and societal values evolve in unpredictable ways. Science fiction offers glimpses of possible futures, allowing us to anticipate and adapt to the changes that may unfold.

Science fiction can show us how future interstellar distances may render current military command structures obsolete. Light-year communication delays will require starship commanders to make autonomous decisions without guidance from a central authority. Adaptability becomes paramount as unforeseen threats emerge in the vast unknown. Leaders will need to identify effective tactics and discard outdated ones as situations evolve. Creative resource management and unorthodox tactics may be crucial for survival if resupply across interstellar gulfs is impossible. This battlespace demands a new breed of strategist: fiercely independent, adaptable, and capable of forging victory from the very fabric of space itself.

By analyzing fictional conflicts, from interstellar battles to guerrilla uprisings, valuable insights emerge that inform real-world military strategy. The principles of asymmetric warfare employed by the Rebel Alliance in *Star Wars* mirror the challenges faced by contemporary planners seeking to overcome superior forces through unconventional tactics. Science fiction not only illuminates the strategies of insurgents but also compels us to consider the dangers of tyranny and the resourcefulness of those who resist such oppression.

Likewise, the tactics employed by Ender Wiggin in *Ender's Game* vividly illustrate the psychological dimension of warfare. The story highlights the importance of understanding an opponent's psychology, a principle that is equally relevant in real-world military strategy, diplomacy, and negotiation. It also highlights the enduring psychological toll that accompanies war.

Our exploration has not shied away from the ethical dilemmas inherent in military strategy, whether presented in the speculative pages of science fiction or experienced on contemporary battlefields. The moral choices made by characters such as Captain Jean-Luc Picard in *Star Trek* or Rear Admiral William Adama in *Battlestar Galactica* can force us to confront the profound ethical dimensions of war. Both warriors must confront the sacrifice of individuals for the greater good, civilian control over military authority, and the moral costs of war.

Science fiction doesn't shy away from war's brutality. It serves a vital role in stripping away the glamor of war and revealing its grim, costly reality. It exposes the horrors inflicted on individuals and societies, the moral compromises fueled by deceit and betrayal, and the deep-seated fear and anger left in its wake. Take *Ender's Game*, where child soldiers are ruthlessly exploited to commit genocide, or *The Forever War*, which portrays a protracted and bloody conflict between humans and aliens that leaves soldiers alienated from their home planet and loved ones.

Science fiction also warns us of potential future risks. Take, for instance, the integration of artificial intelligence into warfare, as depicted in the works of Arthur C. Clarke and Isaac Asimov. The rise of AI and autonomous systems has opened new frontiers in military strategy, calling for innovative approaches to ethics, law, and decision-making. The lessons drawn from science fiction, with its ability to peer into alternative futures, provide us with invaluable tools as we navigate this unfamiliar terrain.

War is an integral facet of science fiction and, undeniably, an intrinsic part of our own reality. It is a reality we cannot afford to ignore or deny. Rather, it is a challenge we must confront. War revolves around people, people unable to reconcile their differences through peaceful means, people resorting to inflicting pain and suffering to achieve their ends, and those unfortunate souls compelled to endure the consequences. War tests our character and determines our destiny, is a trial of our strength and wisdom, and a barometer of our hope and faith.

Our journey also reveals a crucial truth: violence is not always the answer. Some problems simply defy traditional solutions. This does not mean that individuals should quit trying to address their problems. Even when a complete solution seems elusive, there may be ways to manage or mitigate the problem's impact. In such cases, resilience becomes key. We may need to learn to coexist with challenges and find coping mechanisms to navigate their effects.

This exploration, while concluding, is by no means at its end. The world of strategy is dynamic, ever-evolving, and often unpredictable. As a result, there lies a potential for innovation and progress. I invite you to continue your journey. Whether through reading, cinema, discussions, or even crafting your own strategic narratives, you have the power to contribute to the ongoing discourse on military strategy. Your contributions can shape its future.

So, as we part ways, let us carry forward the knowledge and insights gained from our journey. Armed with the wisdom of science fiction and the lessons of history, we are better equipped to confront tomorrow's challenges and build a more peaceful future.

NOTES

1. WAR

1. This is an adaptation of "Peace is not the absence of conflict, but the ability to cope with conflict by peaceful means" (Reagan 1982).
2. Thucydides, an ancient Athenian historian and general, observed, "The strong do what they have the power to do, and the weak must accept what they have to accept" (Strassler 1996).
3. B. H. Liddell Hart described strategy as "the art of distributing military means to fulfil the ends of policy" (Hart 1954).
4. This concept is discussed in Lykke Jr. (1989).
5. This was originally noted in Clausewitz (1976).
6. This paraphrases the Roman philosopher Seneca's idea that luck occurs when preparation meets opportunity.

2. POWER

1. The U.S. government limits its definition of national power in terms of diplomatic, informational, military, and economic instruments, collectively known as DIME (*U.S. Department of Defense 2013*).
2. *Pirate* comes from the Greek root *peirein* "to attempt or attack," where *mercenary* originated from Latin *mercēnārius*, or "hireling," which originally meant someone who will do anything in exchange for money.

3. ENDS

1. Clausewitz (1976) observes that "war is a mere continuation of politics with other means."
2. *Casus belli* means "occasion for war" and refers to actions or events used to justify war against a society. *Casus foederis*, or "case for the alliance," concerns actions against an ally that warrant war. *Casus iustum*, meaning "just cause," denotes the moral or legal rationale for legitimizing war, often grounded in defense, treaty obligation, or protection of innocents. All three concepts originate in Roman law and diplomatic practice and were later shaped by Christian theology.
3. The concept was originally advanced by Machiavelli (Machiavelli and Bull 2003).
4. This concept was originally proposed by Carlin (2018).
5. This idea, termed 'swaggering' was introduced by Art (1980).

5. MOBILIZATION

1. For more on this concept see (Tzu and Cleary 1988).

2. For more on this subject, see two-time Medal of Honor recipient Major General Smedley D. Butler (Butler 1935). Available at:
   https://hdl.handle.net/2027/inu.32000014248506.

6. DOGS OF WAR

1. The phrase originated in *Julius Caesar*, Act 3, Scene 1, when Mark Antony declares, "…cry 'havoc!' and let slip the dogs of war" (Shakespeare 1996, p597). Klingon General Chang later echoed the line in *Star Trek VI: The Undiscovered Country* (Meyer et al. 1991).
2. The philosophy was memorably stated by Science Officer Spock as, "The needs of the many outweigh the needs of the few" in *Star Trek II: The Wrath of Khan* (Meyer 1982).

7. CONTROLLING VIOLENCE

1. In 1519, Hernán Cortés scuttled his ships off the coast of Mexico, eliminating retreat and forcing his men to commit fully to the conquest of the Aztec Empire, a classic example of the "burn the ships" strategy.
2. The ancient Spartans and Vikings regarded restraint as weakness. Similarly, Japan signed but did not ratify the 1929 Geneva Convention before World War II, believing it conflicted with their cultural values.
3. This book does not attempt a comprehensive discussion of wartime morality but instead offers a starting point for considering the ethical dimensions of warfare. The complexity of moral questions in armed conflict requires deeper examination beyond the scope of this work.

8. MILITARY COMMAND

1. This is related to Clausewitz's concept of the "Center of Gravity" (von Clausewitz 1976).
2. This is an example of switching from an attrition strategy to *absolute war*.
3. U.S. Air Force Colonel John Boyd developed the OODA loop concept. See http://www.danford.net/boyd/essence1.htm
4. This phrase is often attributed to General George S. Patton.
5. For further discussion, see *The OODA Loop and the Half Beat* by Alastair Luft, available at: https://thestrategybridge.org/the-bridge/2020/3/17/the-ooda-loop-and-the-half-beat.
6. For more on communicating with alien races, see N. Drake (2019); Chris Baraniuk (2016); and Snedeker (2022)
7. The saying "A pint of sweat will save a gallon of blood" is often attributed to General George S. Patton.

9. MILITARY POWER

1. In the U.S. military, this concept is referred to as combined arms, the integration of different combat arms, such as infantry, armor, artillery, air support, engineers, and

cyber forces, in a synchronized and simultaneous manner to create effects greater than any single element could achieve alone.

2. Selecting a leader through personal combat is often ineffective. A skilled warrior may enforce discipline within a small group but lack leadership or strategic ability. In battle, a leader may either fight or lead, and doing both well is rarely possible.

10. GEOGRAPHY

1. Jomini discussed this concept in (Jomini, Mendell, and Craighill 1892).
2. Adapted from (Cornish 2023)
3. This scenario illustrates the geometric consequences of territorial expansion. The volume of a sphere is calculated by the formula $V = 4/3\ \pi r^3$. At a radius of 10 parsecs, the controlled space is approximately 4,187 cubic parsecs. Doubling the radius to 20 parsecs increases the volume to approximately 33,493 cubic parsecs, a growth factor of eight. If the military force remains the same size, it now occupies only 12.5 percent of the volume it once defended. Spreading those forces around the new perimeter results in a defensive belt only 0.9 parsecs thick. This demonstrates the severe strategic penalties of overextension without proportional buildup.

11. MANEUVER

1. Under the Geneva and Hague Conventions, disguises are lawful only as ruses, such as camouflage, decoys, or misleading movements. Wearing the enemy's uniform or insignia is permitted solely for deception before combat; using it during battle or to gain advantage constitutes *perfidy* and is prohibited as a war crime. Similarly, disguising as members of protected classes, including civilians, medics, or neutral personnel, is forbidden because it violates the protections guaranteed under international humanitarian law.
2. The term comes from the German World War I military tactic known as Kesselschlacht.

12. PSYCHOLOGY

1. Carl von Clausewitz is credited with creating the concept of *friction of war* (von Clausewitz 1976).
2. U.S. General Dwight D. Eisenhower is often quoted for versions of this phrase. See (Eisenhower 1957).
3. This is attributed to Prussian strategist Helmuth von Moltke the Elder: "No plan of operations extends with any certainty beyond the first encounter with the main enemy forces. Only the layman believes that in the course of a campaign he sees the consistent implementation of an original thought that has been considered in advance in every detail and retained to the end" (von Moltke 1900).
4. This occurred during the French military revolt of 1917, which saw widespread mutinies among exhausted and demoralized soldiers, sparked by horrific trench conditions, catastrophic losses, and disillusionment with incompetent leadership.
5. The phrase "hearts and minds" is attributed to French general Hubert Lyautey.

13. THE DARK ARTS

1. Carl von Clausewitz used the term "fog of war" as a metaphor for uncertainty in war (von Clausewitz 1976).
2. From (Menosky and LaZebnik 1991) and discussed in (Bogost 2014).
3. See the Coventry Argument in (VanPutte 2016).
4. This is originally stated by Sun Tzu (Tzu and Cleary 1988).
5. A *paper tiger* is a person or organization that appears threatening but is actually ineffective or powerless.
6. A *poison pill* is a defensive tactic that deters conquerors by presenting oneself as unattractive or undesirable.
7. For more on the indirect approach, see (Hart 1954)

14. LOGISTICS

1. German Field Marshal Erwin Rommel is credited with saying, "The battle is fought and decided by the quartermasters, long before the shooting begins."
2. A 100,000-person army requires 18,000,000 pounds of food and water for that period, and 225 modern-day semi-trucks to haul it. They would also need drivers, fuel, mechanics, and another 6 semis to transport food for these additional people.
3. The Allied invasion of Normandy in 1944 (D-Day) required 810 tons of logistics for each of the 14,000-to 16,000-person divisions. This included fuel, food, and ammunition for an advancing unit in heavy combat. This is approximately 110 pounds per person per day.
4. Ancient warriors discovered that horses and oxen reduced the load they had to carry themselves but created new logistical challenges. A horse requires six times as much food as a human. This dependency increased the demand for foraging and transport, limiting military campaigns to seasons and regions where they could feed their animals (Leighton 1998).
5. A Q-ship is a heavily armed vessel disguised as a harmless merchant or civilian ship, used primarily to lure enemy warships or raiders into attacking what appears to be an unarmed target, only to reveal its true self at close range.
6. On Earth, RMAs include the rise of organized warfare in settlements enabled by agriculture, fortifications, and bronze; mass infantry with iron weapons and disciplined drill; professional standing armies supported by administration and engineering; feudal cavalry with armor, stirrups, and the warhorse; the gunpowder revolution; mass mobilization through conscription and logistics; industrialized warfare driven by railroads, telegraphs, rifled artillery, and machine guns; mechanized and air warfare with tanks, aircraft, radios, and combined arms; atomic weapons; and the shift toward information-centered warfare with precision weapons, satellites, digital networks, cyber, and space systems.
7. The phrase "death by a thousand cuts" originates from the Lingchi, a brutal Chinese execution method that involved slowly slicing condemned individuals over days. It metaphorically describes situations where repeated, minor injuries or insults lead to physical, emotional, or psychological collapse.

17. OFFENSE

1. This was proposed by Prussian military theorist Carl von Clausewitz, who said, "Every attack loses impetus as it progresses" (von Clausewitz 1976).
2. This concept is often attributed to Bertrand Russel, as "War does not determine who is right, only who is left."
3. This is known as the Fabian Strategy, after Quintus Fabius Maximus Verrucosus, a Roman dictator who used it to outmaneuver the Carthaginian general Hannibal during the Second Punic War (218-201 BC).
4. Sun Tzu advised offering foes a "golden bridge" to retreat across, noting that cornered enemies fight like desperate tigers (Tzu and Cleary 1988).

18. DEFENSE

1. For more on caution when pursuing an opponent, see (Tzu and Cleary 1988).

19. INTIMIDATION

1. This concept is expressed as winning without fighting, in (Tzu and Cleary 1988).

20. SUBVERSION

1. Guerrilla is derived from the Spanish word for "small war."
2. Praetorianism refers to the influence of the Roman Praetorian Guard, the elite bodyguards and intelligence agents serving ancient Roman emperors. Their proximity to power enabled them to manipulate imperial succession, conduct assassinations, and suppress opposition. The term describes military interference in civilian politics.
3. This is a paraphrase from Machiavelli (Machiavelli and Bull 2003).
4. For more on this, see (Ganor 2002).

21. POLITICAL STRATEGIES

1. Some may include détente (French for "relaxation"), a temporary easing of hostilities without a formal alliance, and entente (French for "agreement"), an informal understanding to consult or coordinate during times of crisis.
2. The assassination of Archduke Franz Ferdinand of Austria-Hungary in Sarajevo in 1914 set off a chain reaction among Europe's intertwined alliances, drawing more than thirty nations into what became World War I. A regional crisis escalated into a global conflict that, by 1918, had claimed over sixteen million lives and left empires in ruins, reshaping the course of the twentieth century.
3. This occurred in 1940 when Germany invaded the Netherlands and Belgium, bypassing much of France's Maginot Line.

GLOSSARY

1. Based on (Department of the Army 2017).
2. Based on (Department of the Army 2017).

ADDITIONAL READINGS

Alexander, Bevin. *How Wars are Won: The 13 Rules of War*. The Crown Publishing Group, 2007. This provides thirteen rules of war that have survived the ages.

Bassford, Christopher. *Policy, Politics, War, and Military Strategy*. An excellent read on the use of organized violence in the pursuit of political objectives. Available at http://www.clausewitz.com/readings/Bassford/StrategyDraft.

Butler, Smedley. *War is a Racket*. 1935. New York: Round Table Press, Inc. This provides insights into the military-industrial complex that profits during war at the expense of the warriors.

Clausewitz, Carl von. 1976. *On War*. Edited by Michael Howard and Peter Paret. Revised. Princeton University Press. A student of the Napoleonic War, many in the West consider Clausewitz the most important strategist on European military theory.

Dailey, Brian and Parker, Patrick. *Soviet Strategic Deception*. 1987. Lexington Books. An excellent tome on how the Soviet Union conducted deception operations, offering insights into how culture shapes a different mindset toward deception in political and military operations.

Dewar, Michael. 1989. *The Art of Deception in Warfare*. David & Charles Publishing. The definitive guide to the use of deception in warfare.

Dunnigan, James F. 2003. *How to Make War: A Comprehensive Guide to Modern Warfare in the Twenty-first Century*. 4th ed. Edition. William Morrow Paperbacks. A guide to the creation of war games with extensive discussions on all aspects of modern warfare tactics and technologies.

Evans, David (Air Marshal). *War: A Matter of Principles*. 1997. Palgrave Macmillan. An excellent look at the principles of war of the United States and the United Kingdom, and how they differ across countries and services.

Grossman, Dave. *On Killing: The Psychological Cost of Learning to Kill in War and Society*. Little, Brown and Company; 1st edition (October 1, 1995). An excellent guide to the psychology of how governments motivate soldiers to kill, and the cost of war to soldiers and society.

Huntington, Samuel. 1981. *The Soldier and the State: The Theory and Politics of Civil–Military Relations*. Belknap Press. The guide to understanding the military mind and how it interacts with other aspects of the state.

Jomini, Antoine Henri. *The Art of War: Strategy & Tactics from the Age of Horse & Musket*, Aug 13, 2010. As one of the first great European military strategists, Jomini developed fundamental principles of war based on his study of Napoleon, available at https://www.gutenberg.org/files/13549/13549-h/13549-h.htm

Keegan, John. 1993. *A History of War*. New York, NY: Knopf Doubleday Publishing Group. This provides an interesting and in-depth discussion of the evolution of military strategy and tactics, with a focus on how culture influences and drives warfare.

Liddell Hart, Basil Henry (Sir), *Strategy*. (2nd rev. ed.). New York, N.Y., U.S.A.: Meridian. Hart was a combat veteran of the First World War and a British military historian. He was a severe critic of other military strategists, especially of Clausewitz. His writings provide an alternative view to earlier writings, including defining and advocating the use of an indirect approach.

Machiavelli, Niccolo, and George Bull. 2003. *The Prince*. London, England: Penguin Classics. Machiavelli was a strategist in the fields of unrestricted warfare and politics. Though often regarded primarily as a treatise on politics, Machiavelli also addresses significant points of military theory and is considered one of the cornerstones of military thought.

Mao Tse-tung, (U.S. Marine Corps). *Mao Tse-tung on Guerrilla Warfare*. FMFRP 12-18. 1989. This is an English translation of Mao Tse-tung's classic guide on the planning and conduct of guerrilla operations. It is available at https://www.marines.mil/Portals/1/Publications/FMFRP%2012-18%20%20Mao%20Tse-tung%20on%20Guerrilla%20Warfare.pdf

Peter Paret, Gordon A. Craig, Felix Gilbert. *Makers of Modern Strategy from Machiavelli to the Nuclear Age*. Princeton University Press. 1986. Excellent text on many of the West's greatest military strategists.

Potter, E.B. *Sea Power*, Annapolis, MD, USNI Press: 1972. Text used by the United States Naval Academy for teaching (the American view) of naval power.

Strange, Susan. <u>What is economic power, and who has it</u>? *International Journal*, 30(2), 207-224. This provides an introduction and analysis of economic power as an instrument of the state.

ADDITIONAL READINGS

The Science of Military Strategy edited by Peng Guangqian, Yao Youth, 1st Edition, Military Science Publishing House, 2005, Beijing China. An in-depth look into military strategy as seen by China's People's Liberation Army.

Tzu, Sun (translated by) Cleary, Thomas. 1988. *The Art of War*. Boston, MA: Shambhala Publications, Inc. Published between 453-221 B.C.E. and used in the east as the definitive guide to all conflict, from military to business. A must-read for all military strategists.

U.S. Army Doctrine Publication 4-0 published 31 July 2012. This provides a glimpse into U.S. military sustainment principles and logistics.

INDEX OF MAXIMS

1. Power defines possibilities.
2. The means shape the ways.
3. Strategy is an art, not a science.
4. War is a society's ultimate gamble.
5. Geography defines power.
6. Every dependency is a vulnerability.
7. Diversity breeds independence.
8. Politics and warfare often follow wealth and influence.
9. Superior technology gives effective commanders a decisive advantage.
10. Safeguard future options.
11. Timeliness enhances effectiveness.
12. Power demands constant vigilance.
13. Victory is achieved by seizing and maintaining the initiative.
14. Perspective and context shape interpretation.
15. Overreach often incites internal unrest.
16. Communication is the key to understanding and influence.
17. War is an affirmation of a society.
18. Know your enemy. Hide yourself.
19. Coercion may impose obedience but not loyalty.
20. All military operations must advance the ends.
21. Strike when and where an opponent is vulnerable.
22. Reserves provide battlefield flexibility.
23. A battle is often decided before the first round is fired.
24. Unchecked fear breeds ruin.
25. When faced with a no-win situation, change the situation.
26. Greater violence leads to greater resentment.
27. Control and exploit key terrain.
28. Project power to increase influence.
29. Don't strike a foe's home and neglect your own.
30. Troopers are a leader's proxy.
31. Simplicity enables effective decisions.
32. Flexibility is strength; predictability is weakness.
33. Control their minds and you control their actions.
34. Security forces enable initiative and flexibility.
35. Eliminate distractions to find crucial information and opportunities
36. The ignorant and the dead keep secrets.
37. Shrewdness guides wisdom.
38. The essence of deception lies in confirming preconceptions.
39. An indirect path conceals intent.
40. A force may forage, fabricate, store, or receive logistics from others.

CITATIONS

Abnett, Dan. 2005. *Eisenhorn Trilogy*. Black Library.

Abnett, Dan. 2006. *Horus Rising (The Horus Heresy)*. Black Library.

Abrams, J.J. 2009. *Star Trek*. Paramount Pictures - USA.

Abrams, J.J. 2013. *Star Trek into Darkness*. Paramount Pictures - USA.

Abrams, J.J., Lawrence Kasdan, and Michael Arndt. 2015. *Star Wars: Episode VII - The Force Awakens*. United States: Walt Disney Studios Motion Pictures.

Adams, Douglas. 1979. *The Hitch-Hiker's Guide to the Galaxy*. New York, NY: Del Rey.

Anderson, Poul. 1957. Among Thieves. *Astounding Science Fiction*, June.

Art, Robert J. 1980. To What Ends Military Power? *International Security* 4 (3 (Spring 1980)):3–35.

Asimov, Isaac. 1950. *I, Robot*. New York, NY: Gnome Press.

Asimov, Isaac. 1951. *Foundation*. Garden City, NY: Doubleday.

Asimov, Isaac. 1953. *Second Foundation*. Garden City, NY: Doubleday.

Axe, David. 2020. Bad news: Russia has no strategy to stop an alien invasion. *The National Interest*. (November 5)

Baird, Stuart. 2002. *Star Trek: Nemesis*. USA: Paramount Pictures.

Berezhnoy, Sergey. 2013. Russian Space Troops Are Not Prepared for Battle with Aliens, Says Official. *The Independent*. See www.the-independent.com/news/world/europe/russian-space-troops-are-not-prepared-for-battle-with-aliens-says-official-8856460.html.

Berman, Rick, Brannon Braga, Ronald D. Moore, and Jonathan Frakes. 2009. *Star Trek, First Contact*. United States: Hollywood, CA: Paramount Home Entertainment.

Berman, Rick, Michael Piller, and Jeri Taylor. 1995. Star Trek: Voyager, "Caretaker", (Season 1, Episodes 1 and 2). United States.

Berman, Rick, Michael Piller, Jeri Taylor, and Ira Steven Behr. 1994. Star Trek: Deep Space Nine, "The Maquis, Part II", (Season 2, Episode 21). United States.

Biller, Kenneth. 1997. Star Trek: Voyager, "Unity", (Season 3, Episode 17). United States.

Bogost, Ian. 2014. Shaka, When the Walls Fell. *The Atlantic*, June 18. https://www.theatlantic.com/entertainment/archive/2014/06/star-trek-tng-and-the-limits-of-language-shaka-when-the-walls-fell/372107/.

Boulding, Kenneth. 1962. *Conflict and Defense: A General Theory*. New York, NY: Harper.

Bowman, Rob. 1987. Star Trek: The Next Generation: "The Battle", (Season 1, Episode 9).

Bowman, Rob. 1988. Star Trek: The Next Generation, "Heart of Glory" (Season 1, Episode 19):

Brancato, John, and Michael Ferris. 2009. *Terminator Salvation*. United States: Warner Home Video.

Bray, Adam, Cole Horton, and Tricia Barr. 2017. *Star Wars: The Visual Encyclopedia*. Illustrated Edition. DK Publishing.

Bujold, Lois McMaster. 2000. *Shards of Honor (Vorkosigan Saga Volume 1)*. Framingham, MA: NESFA Press.

CITATIONS

Bungie. 2004. *Halo 2*. Redmond, WA: *Microsoft Game Studios*.

Bungie. 2010. Halo: Reach. *Microsoft Game Studios*.

Butler, Smedley D. 1935. *War Is a Racket*. New York, NY: Round Table Press, Inc.

Cameron, James. 1984. *The Terminator*. United States: Orion Pictures.

Cameron, James. 1986. *Aliens*. United States: Twentieth Century Fox.

Cameron, James. 2009. *Avatar*. United States: Twentieth Century Fox.

Cameron, James, and William Wisher. 1991. *Terminator 2: Judgement Day*. United States: Artisan Home Entertainment.

Campbell, Jack. 2006. *The Lost Fleet: Dauntless*. Ace Books.

Campbell, Jack. 2007. *The Lost Fleet: Fearless*. Ace Books.

Campbell, Jack. 2011. *The Lost Fleet: Victorious*. Ace Books.

Campbell, Jack. 2013. *The Lost Fleet: Invincible*. Ace Books.

Campbell, Jack. 2015. *The Lost Fleet: Beyond the Frontier: Leviathan*. Ace Books.

Card, Orson Scott. 1999. *Ender's Game*. London: Orbit.

Card, Orson Scott, and John Harris. 1991. *Xenocide*. New York, NY: Tor.

Card, Orson Scott, and Aaron Johnston. 2014. *Earth Afire: The First Formic War*. New York, NY: Tor.

Card, Orson Scott, and Aaron Johnston. 2016. *The Swarm, Second Formic War Volume 1*. New York: Tor.

Carlin, Dan. 2018. Hardcore History 62 – Supernova in the East I. *Https://Www.Dancarlin.Com/Product/Hardcore-History-62-Supernova-in-the-East-i/*.

Cherryh, C.J. 1986. The Scapegoat. In *Body Armor/2000*, ed. Joe Haldeman. Ace.

Chris Baraniuk. 2016. What Would Happen If Aliens Contacted Earth? *BBC Future*, February 23. https://www.bbc.com/future/article/20160223-what-would-happen-if-aliens-contacted-earth/.

Clarke, Arthur C. 1951. Superiority. *The Magazine of Fantasy & Science Fiction*, August.

Collins, Suzanne. 2008. *The Hunger Games*. Scholastic Press.

Contner, James A, and Maria and Andre Jacquemetton. 2002. Star Trek: Enterprise, "Dear Doctor", (Season 1, Episode 13). UPN.

Corey, James S.A. 2011. *The Expanse: Leviathan Wakes*. New York, NY: Orbit/Hachette Book Group.

Corey, James S.A. 2012. *The Expanse: Caliban's War*. New York, NY: Orbit.

Cornish, Neil J. 2023. What Is a Lagrange Point? https://science.nasa.gov/resource/what-is-a-lagrange-point/.

Coyle, Paul Robert. 1995. Star Trek: Voyager, "State of Flux", (Season 1, Episode 11). Paramount Television.

Daniels, Marc. 1967. Star Trek, "The Doomsday Machine" (Season 2, Episode 6). NBC.

Department of the Army. 2003. *Field Manual 6-0, Mission Command: Command and Control of Army Forces*. Washington, D.C.

Devlin, Dean, and Roland Emmerich. 1994. *Stargate*. United States: MGM/UA Distribution Co.

Dewar, Michael. 1989. *The Art of Deception in Warfare*. David & Charles Publishing.

Diamond, Jared M. 1999. *Guns, Germs, and Steel: The Fates of Human Societies*. New York, NY: W.W. Norton & Co.

Drake, David. 2009. *The Complete Hammer's Slammers: The Irresistible Force*. Baen Books.

Drake, Nadia. 2019. How Would Humans React to the Discovery of Alien Life? *National*

384

Geographic, February. https://www.nationalgeographic.com/science/article/how-would-people-react-alien-life-discovery-aaas-space-science.

Echevarria, René. 1992. Star Trek: The Next Generation, "I, Borg", (Season 5, Episode 23).

Edwards, Gareth. 2016. *Rogue One: A Star Wars Story*. Walt Disney Studios Motion Pictures.

Eisenhower, Dwight D. 1957. *Remarks at the National Defense Executive Reserve Conference*. Washington, D.C. https://hdl.handle.net/2027/miua.4728417.1957.001?urlappend=%3Bseq=858.

Eisner, Breck. 2019. The Expanse: "Jetsam", (Season 4, Episode 2).

Eisner, Breck, Ty Franck, Daniel Abraham, and Mark Fergus. 2017. The Expanse: "Doors & Corners", (Season 2, Episode 2). United States: Boom! Studios.

Emmerich, Roland. 1996. *Independence Day*. United States: Twentieth Century Fox.

Fergus, Mark, Hawk Ostby, and Terry McDonough. 2015. The Expanse: "The Big Empty" (Season 1, Episode 2).

Fink, Kenneth, Ty Franck, Daniel Abraham, and Mark Fergus. 2017. The Expanse: "The Seventh Man" (Season 2, Episode 17).

Fontana, D.C. 1967. Star Trek: "Journey to Babel", (Season 2, Episode 10). United States.

Fontana, D.C. 1968. Star Trek, "The Enterprise Incident", (Season 3, Episode 4) .

Franck, Ty, Daniel Abraham, and Mark Fergus. 2017. The Expanse: "Static", (Season 2, Episode 3).

Freudenthal, Thor. 2017. The Expanse: "Caliban's War", (Season 2, Episode 13).

Futamura, Hideki. 2010. *Halo: Legends, Origins*. USA: Warner Brothers.

Ganor, Boaz. 2002. Defining Terrorism: Is One Man's Terrorist Another Man's Freedom Fighter? *Police Practice and Research: An International Journal* 3 ((4)):287–304.

Gilroy, Dan. 2022. *Andor*: "Aldhani", (Season 1, Episode 4). Disney+, September 28, 2022.

Goldsman, Akiva. 2017. Star Trek: Discovery, "Context Is for Kings", (Season 1, Episode 3).

Grabiak, Marita. 2005. Battlestar Galactica: "Water" (Season 1, Episode 2).

Gray, Claudia. 2017. *Star Wars: Lost Stars*. Egmont, UK,: Disney-Lucas Press.

Grossman, Dave. 1995. *On Killing: The Psychological Cost of Learning to Kill in War*. 1st Edition. Boston: Little, Brown, and Company.

Gunn, Eileen. 2014. How America's Leading Science Fiction Authors Are Shaping Your Future. *Smithsonian Magazine*, May.

Haight, Wanda M., Gregory W. Amos, and Burton Armus. 1989. Star Trek: The Next Generation, "A Matter of Honor", (Season 2, Episode 8). United States.

Haldeman, Joe. 1974. *The Forever War*. New York: Ballantine Books.

Haldeman, Joe, and Ben (editor) Bova. 1972. Hero. *Analog Science Fiction/Science Fact* LXXXIX (4):8.

Hart, B.H. Liddell. 1954. *Strategy: The Indirect Approach*. Third Edition. London: Faber and Faber.

Heinlein, Robert A. 1960. *Starship Troopers*. Hodder & Stoughton Publishing.

Helbing, Todd, Aaron Helbing, Stewart Hendler, Thom Green, and Anna Popplewell. 2012. *Halo 4: Forward Unto Dawn*. United States: Microsoft Corporation.

Herbert, Frank. 1965. *Dune*. Philadelphia: Chilton Books.

Herbert, Frank. 1976. *Children of Dune*. New York, NY: Berkley Publishing Corp.

Hidalgo, Pablo. 2016. *Star Wars Propaganda: A History of Persuasive Art in the Galaxy*.

Harper Design.

Holdstock, Robert, ed. 1978. *Encyclopedia of Science Fiction*. London: Octopus Books.

Jomini, Antoine Henri, G. H. (translated) Mendell, and W. P. Craighill. 1892. *The Art of War*. Philadelphia, PA: J. B. Lippincott Company.

Jones, Raymond F. 1953. Noise Level. *Astounding Science Fiction* #18 (Dec.):p172.

Karpyshyn, Drew. 2007a. *Mass Effect: Revelation (Volume 1)*. New York: Del Rey.

Karpyshyn, Drew ed. 2007b. Mass Effect. Electronic Arts, dev. Bioware .

Kasdan, Jonathan, and Lawrence Kasdan. 2018. *Solo: A Star Wars Story*. United States: Walt Disney Studios Motion Pictures.

Kemper, David. 1989. Star Trek: The Next Generation, "Peak Performance", (Season 2, Episode 21).

Kierkegaard, Søren, and Alexander Dru. 2003. *The Soul of Kierkegaard: Selections from His Journals*. New York, United States: Dover Publications.

Klink, Lisa, and Nicholas Corea. 1995. Star Trek: Deep Space Nine, "Hippocratic Oath", (Season 4, Episode 3). USA: Paramount Television.

Kolbe, Winrich. 1993. Star Trek: The Next Generation, "Rightful Heir" (Season 6, Episode 23):

Krauss, Lawrence M. 1995. *The Physics of Star Trek. Basic Books*. New York, NY: Basic Books.

Lambert, Hallie. 2021. The Expanse: "Tribes", (Season 5, Episode 6).

Larson, B.V. 2015. *Battle Cruiser (Lost Colonies Trilogy)*. CreateSpace Independent Publishing Platform.

Larson, Glen A. 2009. Battlestar Galactica. The Complete Series. Universal City, CA: Universal Studios.

Lee, Steward, George Lucas, and Matt Michnovetz. 2011. Star Wars: The Clone Wars, "Citadel Rescue", (Season 3 Episode 20). USA.

Leighton, Richard M. 1998. Logistics. Edited by The Editors of Encyclopedia Britannica. *Encyclopedia Britannica*. Encyclopedia Britannica, Inc.

Lewis, Danny. 2016. Reagan and Gorbachev Agreed to Pause the Cold War in Case of an Alien Invasion. *Smithsonian Magazine*, September.

Lin, Justin, Simon Pegg, and Doug Jung. 2016. *Star Trek Beyond*. Paramount Pictures - USA.

Lucas, George. 1977. *Star Wars - Episode IV: A New Hope*. Twentieth Century Fox.

Lucas, George. 1980. *Star Wars - Episode V: The Empire Strikes Back*. Twentieth Century Fox.

Lucas, George. 1983. *Star Wars - Episode VI: Return of the Jedi*. Twentieth Century Fox.

Lucas, George. 1999. *Star Wars - Episode I: The Phantom Menace*. USA: Twentieth Century Fox Home Entertainment.

Lucas, George. 2002. *Star Wars - Episode II: Attack of the Clones*. Full scree. Beverly Hills, Calif.: Twentieth Century Fox.

Lucas, George. 2005. *Star Wars - Episode III: Revenge of the Sith*. United States: 20th Century Fox Home Entertainment.

Lykke Jr, Arthur F. 1989. Defining Military Strategy. *Military Review* 69 (5):2–8.

Lynch, David. 1984. *Dune*. USA: Universal Pictures.

MacArthur, Douglas. 1962. Duty, Honor, Country. *United States Military Academy*. West

Point, NY: American Rhetoric. See www.americanrhetoric.com/speeches/douglas-macarthurthayeraward.html.

Machiavelli, Niccolo, and George Bull. 2003. *The Prince*. London, England: Penguin Classics.

Manners, Kim, and Hannah Louise Shearer. 1988. Star Trek: The Next Generation, "When the Bough Breaks", (Season 1, Episode 17). USA.

McEveety, Vincent, and Paul Schneider. 1966. Star Trek: "Balance of Terror" (Season 1, Episode 14).

Mehrabian, Albert. 1981. *Silent Message: Implicit Communications of Emotions and Attitudes*. Belmont, CA: Wadsworth.

Melching, Steven. 2008. Star Wars: The Clone Wars, "Rising Malevolence", (Season 1, Episode 2). United States.

Menosky, Joe, and Phillip LaZebnik. 1991. Star Trek: The Next Generation, "Darmok", (Season 2, Episode 5). United States.

Meyer, Nicholas. 1982. *Star Trek II: The Wrath of Khan*. Paramount Pictures - USA.

Meyer, Nicholas, Denny Martin Flinn, Leonard Nimoy, Lawrence Konner, and Mark Rosenthal. 1991. Star Trek VI: The Undiscovered Country. United States: Paramount Pictures.

Mimica-Gezzan, Sergio, Paul T Scheuring, Mike Colter, Steven Waddington, and Christina Chong. 2015. *Halo: Nightfall*. United States: Microsoft.

Moore, Ronald D. 1990. Star Trek: The Next Generation, "The Defector", (Season 3 Episode 10). USA: Paramount Domestic Television.

Moore, Ronald D. 1994. Star Trek: The Next Generation, "Journey's End", (Season 7, Episode 20).

Moore, Ronald D. 2004. *Battlestar Galactica: "33"*, (Season 1, Episode 1). Sci-Fi Channel.

Morgan, Glen, and James Wong. 1995. *Space Above and Beyond*. United States: Hard Eight Pictures, 20th Century Fox Television.

N. Gregory Mankiw. 2020. *Principles of Economics*. 10th ed. Cengage Learning.

Newland, John, Gene L. Coon, and Gene Roddenberry. 1967. Star Trek: "Errand of Mercy" (Season 1, Episode 26).

Nylund, Eric S. 2001. *Halo: The Fall of Reach*. New York, NY: Simon & Schuster.

Nylund, Eric S. 2003. *Halo: First Strike*. New York, NY: Random House.

Okuda, Michael, Denise Okuda, and Doug Drexler. 1999. *The Star Trek Encyclopedia: A Reference Guide to the Future*. New York: Pocket Books.

Olmos, Edward James. 2009. Battlestar Galactica: The Plan. United States: Universal Pictures.

Orwell, George. 1949. *Nineteen Eighty-Four: A Novel*. New York: Signet Classic.

Peeples, Samuel A. 1966. Star Trek: "Where No Man Has Gone Before", (Season 1, Episode 3).

Pevney, Joseph, Robert Hamner, Gene L. Coon, and Gene Roddenberry. 1967. Star Trek: "A Taste of Armageddon" (Season 1, Episode 23). United States.

Pevney, Joseph, Gene Roddenberry, and David P. Harmon. 1967. Star Trek: "The Deadly Years" (Season 2, Episode 12).

Piller, Michael. 1990. Star Trek: The Next Generation, "The Best of Both Worlds", (Season 3, Episode 26 and Season 4, Episode 1).

Reagan, Ronald. 1982. *Address at Commencement Exercises at Eureka College in Illinois*.

CITATIONS

Reeves-Stevens, Judith, and Garfield Reeves-Stevens. 2005. Star Trek: Enterprise, "Divergence", (Season 4, Episode 16). United States: UPN.

Relic Entertainment. 2004. Warhammer 40,000: Dawn of War. THQ.

Robinson, Carla. 2005. Battlestar Galactica, "Colonial Day", (Season 1, Episode 11). United States.

Roddenberry, Gene. 1965. Star Trek: "The Cage", (Season 1, Episode 1). United States.

Roddenberry, Gene. 1969. Star Trek: "The Savage Curtain" (Season 3, Episode 22). Desilu Productions.

Roddenberry, Gene (creator), Ira Steven Behr, and Robert Hewitt Wolfe. 1993. Star Trek: Deep Space Nine, "To the Death", (Season 4, Episode 22).

Roddenberry, Gene, and Dorothy C. Fontana. 1966. Star Trek, "Charlie X", (Season 1, Episode 2). United States: NBCUniversal Television Distribution.

Rothstein, Richard, Christopher Leitch, and Dean Devlin. 2004. *Universal Soldier*. United States: Lions Gate Home Entertainment.

Saberhagen, Fred. 2003. *Berserker Prime*. Tor Books.

Sargent, Joseph (director), and Jerry (writer) Sohl. 1966. Star Trek: "The Corbomite Maneuver" (Season 1. Episode 10). USA.

Scalzi, John. 2005. *Old Man's War*. Tor Books.

Scheerer, Robert. 1992. Star Trek: The Next Generation, "Outcasts", (Season 5, Episode 17).

Schelling, Thomas C. 1966. *Arms and Influence*. Yale University Press.

Scott, Ridley. 1979. *Alien*. 20th Century Fox.

Seabury, Paul, and Angelo Codevilla. 1990. *War: Ends and Means*. NewYork, NY: Basic Books.

Shakespeare, William. 1996. *The Complete Works of William Shakespeare*. Wordsworth Edition.

Shatner, William (director), Gene Roddenberry, William Shatner, and Harve Bennett. 1989. *Star Trek V: The Final Frontier*. USA: Paramount Pictures.

Shearman, Robert, and Joe Ahearne. 2005. Doctor Who, "Dalek", (Season 1, Episode 6). BBC.

Snedeker, Claire E. 2022. Oh, If I Could Talk to the Aliens. *How to Talk to Extraterrestrials*, March 10. https://news.harvard.edu/gazette/story/2022/03/how-to-talk-to-extraterrestrials/.

Straczynski, Michael J. 1995a. Babylon 5: "And Now for a Word", (Season 2, Episode 15). USA: TNT.

Straczynski, Michael J. 1995b. Babylon 5: "The Long, Twilight Struggle", (Season 2, Episode 20). USA: TNT.

Straczynski, Michael J. 1997. Babylon 5: "Atonement", (Season 4, Episode 9). USA: TNT.

Straczynski, Michael J. 1998a. *Babylon 5: "In the Beginning."* USA: Babylonian Productions.

Straczynski, Michael J. 1998b. Babylon 5: "The Paragon of Animals", (Season 5, Episode 3). USA: TNT.

Strassler, Robert B., ed. 1996. *The Landmark Thucydides*. New York, NY: A Touchstone Book.

Sussman, Mike. 1996. Star Trek: Voyager, "The Swarm", (Season 3, Episode 4).

Sussman, Mike, and Phyllis Strong. 2003. Star Trek: Enterprise, "Regeneration", (Season 2, Episode 23). United States.

Tapping, Amanda (Director), Michael (Director) Shanks, and Peter (Director) DeLuise. 2020. Stargate SG-1: The Complete Series.

Taylor, Michael. 2007. Battlestar Galactica: Razor. Universal Studios Home Entertainment.

Thompson, Bradley, and David Weddle. 1999. Star Trek: Deep Space Nine. "Extreme Measures". (Season 7, Episode 23). United States: UPN.

Tzu, Sun, and Thomas (translated by) Cleary. 1988. *The Art of War*. Boston, MA: Shambhala Publications, Inc.

United States Space Command. 2023. About U.S. Space Command. *U.S. Department of Defense: See Www.Spacecom.Mil/About-Us*.

U.S. Department of Defense. *Joint Publication 1: Doctrine for the Armed Forces of the United States*. Washington, DC: Joint Chiefs of Staff, 25 March 2013, incorporating Change 1, 12 July 2017.

U.S. Department of Defense. *Joint Publication 3-0: Joint Operations*. Washington, DC: Joint Chiefs of Staff, 17 January 2017, incorporating Change 1, 22 October 2018.

VanPutte, Michael A. 2016. *Walking Wounded: Inside the US Cyberwar Machine*. Createspace Publishing.

Vauban, Sebastien Le Prestre de, and George A. (translated) Rothrock. 1740. *A Manual of Siegecraft and Fortification*. reprint (1968). Ann Arbor, Michigan: University of Michigan Press.

Vejar, Mike. 2002. Star Trek: Enterprise, "Marauders" (Season 2, Episode 6). United States: Paramount Network Television.

Verhoeven, Paul. 1997. *Starship Troopers*. United States: Sony Pictures.

von Clausewitz, Carl. 1976. *On War*. Edited by Michael Howard and Peter Paret. Revised. Princeton University Press.

von Moltke, Helmuth. 1900. *Moltkes Militärische Werke: II. Die Thätigkeit Als Chef Des Generalstabes Der Armee Im Frieden. (Moltke's Military Works: II. Activity as Chief of the Army General Staff in Peacetime) Zweiter Theil (Second Part), Aufsatz Vom Jahre 1871 Ueber Strategie (Article from 1871 on Strategy)*. Berlin, Germany: Ernst Siegfried Mittler und Sohn.

Weber, David. 1994. *The Short Victorious War (Honor Harrington #3)*. New York, NY: Baen Books.

Weber, David, and David B Mattingly. 2006. *In Fury Born*. Riverdale, NY: Baen Books.

Wells, H.G. 1898. *The War of the Worlds*. William Heinemann (UK).

Wells, H.G. 1901. *The First Men in the Moon*. George Newnes.

Wendig, Chuck. 2017. *Empire's End: Aftermath (Star Wars: The Aftermath Trilogy)* . Del Rey.

Whedon, Joss. 2005. *Serenity*. United States: Universal Pictures.

Whedon, Joss. 2014. Firefly. The Complete Series. Twentieth Century Fox Home Entertainment, Inc.

Wilber, Carey, Gene L. Coon, and Carey Wilber. 1967. Star Trek, "Space Seed", (Season 1, Episode 22). United States.

Wolfe, Robert Hewitt. 1994. Star Trek: Deep Space Nine, "The Wire", (Season 2, Episode 22). Paramount Domestic Television.

Wright, Brad, Robert C. Cooper, Joseph Mallozzi, Paul Mullie, Martin Gero, and Carl

Binder. 2004. Stargate Atlantis. MGM Worldwide Television Distribution, Sony Pictures Television (2005-2006).

Zahn, Timothy. 1991. *Heir to the Empire (Star Wars: The Thrawn Trilogy, Vol. 1)*. Bantam Spectra.

Zahn, Timothy. 1993. *The Last Command (Star Wars: The Thrawn Trilogy)*. Bantam Spectra.

Zahn, Timothy. 2017. *Star Wars: Thrawn*. New York: Century.

Zahn, Timothy. 2021. *Thrawn Ascendency: Greater Good*. Del Rey.

Zhang, Stephen Chen. 2021. China's Military Uses AI to Track Rapidly Increasing UFOs, Report Says. *South China Morning Post*, June.

Zicree, Marc Scott. 1994. Babylon 5: "Survivors" (Season 1, Episode 11). USA.

GLOSSARY

The following terms appear throughout this book.

Absolute war (also known as unlimited war): War without moderation or compromise. **See terrorism.**

Active defense (also known as a spoiling attack): A limited attack to destroy, disable, or destabilize forces threatening an imminent attack.

Active sensor: A detection device that emits energy that is often reflected off objects back to a sensor.

Agent: A person who performs an action on behalf of someone else.

Alliance (also known as a pact, agreement, covenant, treaty, or federation): A formal agreement between societies for mutual self-interest. See **coalition.**

Annex: The transition of land from the control of one entity to another, without removing the existing leadership. See **conquest, cession, colonization.**

Annihilation strategy: A strategy that focuses on the destruction of an opponent's military.

Area of influence: The area a commander can directly influence.

Area of interest: The area of concern to the commander, including the **areas of influence,** adjacent areas, and enemy territory.

Area of operation: The area where military forces operate, which may include enemy territory.

Armistice: A formal and permanent agreement to cease all military operations.

Asymmetric effects: Effects generated by means and/or in ways that are disproportional to their effort.

Asymmetric warfare: A conflict between belligerents with dissimilar, and often disproportionate, means and/or ways.

Attrition: The act of using sustained means to gradually reduce an opponent's military strength or effectiveness.

Authority: The set of socially or politically legitimized powers that is granted to an individual or group to exercise control or influence over people, organizations, or institutions.

Auxiliary: Non-combat support vessels or units. Sometimes refers to secondary warfighters from distant provinces or conquered territories.

Balance of power: A situation where societies have roughly equal power, deterring them from attacking or threatening each other.

Battle: The attempt to achieve an objective through military force.

Battlespace: The environment that directly or indirectly affects combat operations and depends on the range of weapons and maneuver of forces; the contested geography that includes the **area of operation, area of influence,** and **area of interest.**

Beaten Zone: The area where the fire from two or more weapons overlaps and can strike the same target.

GLOSSARY

Bingo: The point where a force must return to resupply or risk having insufficient supplies to return.

Blockade: The act of isolating a region to compel its surrender or weaken it prior to an invasion. See **siege**.

Blocking position: A defensive posture to stop, delay, or redirect an advancing enemy by denying access to key terrain or routes.

Booby trap: A device designed to kill, harm, or surprise an unsuspecting person who disturbs an apparently harmless object or performs a presumably safe act.

Bureaucrat: A government official who performs administrative and management functions.

Bypass: an offensive operation that deliberately avoids heavily fortified or unimportant forces or regions.

Cache: A hidden stockpile of supplies.

Campaign: Several related military operations aimed at achieving a particular military objective.

Cannibalize: The practice of extracting usable parts from damaged or non-functional equipment to maintain other equipment.

Ceasefire: A negotiated agreement to cease hostilities, which often includes other peace-making actions such as setting up demilitarized zones.

Cessation of hostilities: A provisional agreement to cease fighting, devoid of a political resolution.

Cession: The action of giving the right to territory to another, such as in a sale or treaty. See **conquest, annex, colonization**.

Chain-ganging: A situation whereby an alliance unintentionally or intentionally pulls its members into a conflict when one member of the alliance goes to war.

Clandestine: An operation whose existence must remain a secret. See **covert**.

Coalition: A temporary alliance to perform some action. See **alliance**.

Coercion: To encourage someone to perform some act using threats of punishment.

Cohort: A military unit whose members are united by common characteristics or goals.

Collateral damage: Civilian casualties and damage to civilian objects that occur incidentally when military forces target a military objective or cause.

Collective punishment: Penalizing a group for actions perpetrated by a member.

Colonization: Taking control and then economically exploiting a society. See **conquest, annex, cession**.

Combat ineffective: A condition in which a military force can no longer perform its assigned mission or pose a viable threat, typically because of significant losses in personnel, equipment, morale, leadership, or logistical support.

Command: The authority and responsibility a leader exercises over subordinates based on their rank or position. [1]

Commander's intent: The purpose and desired outcome of an **operation** that a commander seeks to achieve, independent of the specific **ways** employed.

Commitment: An obligation by an individual or group to a cause. See **dedication**.

Comparative advantage: The ability of one group to create an effect that another group cannot reproduce or effectively counter.

Compellence: Coercing an opponent to change its behavior. See **deterrence**.

Conditional surrender: Involves all parties negotiating the terms of a capitulation. See **unconditional surrender.**

Confirmation bias: A tendency to only credit information that agrees with a preconceived idea or concept, and to discount information that disagrees with this idea or concept.

Conquest: A group seizes territory by changing who controls the land, such as an invasion that removes a sovereign leader. See **cession, colonization, annex.**

Constraint: A requirement that dictates how an action must occur, such as the start time, end time, or the means that may be used. See **restraint.**

Control: The use of means to accomplish a mission.[2]

Cordon: A line of troops or barriers established to control movement into or out of a designated area. See **interdiction.**

Counterattack: An offensive maneuver in response to an enemy's attack to regain lost ground, destroy the attacker, or seize control of the situation.

Counterintelligence: Means and ways used to identify, prevent, and counteract espionage attempts.

Coup d'état (sometimes simply called *coup***):** The violent overthrow or change of an existing government.

Covering force: A military unit operating apart from the main body to intercept, delay, disorganize, and deceive the enemy. See **rear-guard.**

Covert: An operation whose sponsor must be kept secret. See **clandestine.**

Critical components: A society's essential and interdependent elements that shape, sustain, and enable the various **forms of power** necessary to function.

Cruiser: A vessel that can operate independently in distant regions.

Culmination point: The location where an offensive has exhausted its ability to continue because of supply problems, the opposing force, or the need for rest.

Cultural power: A society's intentional or unintentional use of its ideas, achievements, art, literature, or language to inspire and influence others.

Culture: The beliefs, values, and behaviors of a particular society that shape decisions and influence actions.

Culpable individual: Someone believed to support or perform heinous acts.

Decapitation: Operations involving assassinating, kidnapping, incapacitating, or discrediting key individuals through physical, logical, or psychological means.

Deception: Actions taken to conceal or misrepresent the truth.

Decisive engagement: A confrontation that occurs when a military force cannot maneuver or withdraw and must fight to a conclusion with the forces available.

Decisive victory: A military success that resolves a conflict, often by eliminating an opponent's means to resist.

Dedication: The passion or devotion a person or group demonstrates toward fulfilling a commitment.

Demonstration: A show of force to entice an opponent to react in a certain way that does not involve contact with the enemy. See **feint.**

Deterrence: Discouraging an opponent from performing an action by the existence of a credible cost that they believe outweighs the perceived benefits. See **compellence.**

Divergent objective: A situation involving conflicting goals or priorities within or between organizations.

Divide-and-conquer (also known as defeat in detail): A strategy that seeks to defeat an opponent by isolating and destroying its components incrementally rather than confronting the whole at once.

Divine intervention: A miracle that causes or prevents an event.

Doctrine: A group's set of beliefs and principles in the application of violence based on cultural beliefs and experiences.

Droid: A humanoid robot typically known for its versatility and ability to interact with the environment. See **probe.**

Dual-use: Objects with both peaceful and military applications.

End state: The specified situation at the completion of the last phase of a military operation.

Ends: The desired outcomes.

Engagement area: The three-dimensional geographic area where a military force plans to mass its military power.

Enticement: An attempt to lure someone into doing something they know is wrong, involving a benefit.

Envelopment: A maneuver in which attacking forces bypass an opponent's front to strike from the flanks or rear, to encircle and isolate them from support or escape.

Espionage: The practice of covertly collecting information about another actor's intentions, capabilities, or activities.

Ethnophobia: A real or imagined fear or hatred of people of a different ethnicity. See **xenophobia.**

Exchange officer: An officer of one group that is temporarily assigned to, and works for, the group for which they are assigned.

Exhaustion campaign: A strategy that seeks to gradually wear down an opponent's society.

Exoforming: Using engineering and scientific methods to alter a celestial body's atmosphere, temperature, surface conditions, and other features to create a life-supporting environment.

Expeditionary force: An armed force organized to achieve a specific objective in a foreign region (U.S. Department of Defense 2017).

False flag: A deception designed to disguise the identity or origin of the perpetrator.

Feint: A deception to draw enemy forces away from the real objective.

Firepower: The destructive potential of a force's weaponry composed of a combination of weapon volume, accuracy, and effects.

First contact: An initial encounter between two governments, races, or species.

Fog of war: A metaphor for the uncertainty in war.

Force multiplier: Tangible and intangible means or ways that amplify or reduce military power disproportionately to the effort applied.

Force projection: The ability to deploy and sustain military forces in distant regions, enabling a society to exert influence and respond to threats beyond its borders.

Forces: The means available to create military power.

Forlorn hope: A military force assigned a mission with little or no chance of survival.

Form factor: Features related to a component or system's size, shape, and other physical specifications.

Forms of power: The factors that contribute to a society's ability to exert influence and attain its strategic objectives and ends. See **power**.

Friction of war: The unforeseen problems that inevitably occur when executing military operations.

Fundamentalism: The extreme interpretation or enforcement of an ideology, whether religious, nationalistic, cultural, or racial.

Genocide: The deliberate killing of a large portion of a society with the intent of eliminating that society as a whole.

Geopolitics: The study of the intersection of geography, territory, and power.

Geospatial intelligence: The information about a region's physical properties and inhabitants.

Go native: an individual adopts the customs, values, or behaviors of a foreign culture to such an extent that they abandon their original beliefs, identity, or loyalties.

Graft: The use of political power for personal gain.

Grand strategy (also known as national strategy): A state's plan to use military, political, and economic power to obtain its goal.

Gravity well: A region of space around a massive object where the gravitational pull is strong enough to trap objects.

Guerrilla: A member of a small-scale military force using irregular tactics to fight a larger, conventional force.

Guerrilla warfare (also known as a low-intensity conflict or insurgency): Warfare involving a small group of combatants against a larger, more traditional military.

Holding action: An operation designed to prevent an enemy force from moving.

Ideology: A system of social and political beliefs which may have a religious or philosophical basis, and which directly impacts the lifestyle of a society's citizens.

Imperialism: A policy where a society has political, military, or economic control, either formally or informally, over another society.

Indirect approach: A strategy that seeks to undermine an enemy's strength and morale through surprise, maneuver, and disruption rather than a direct confrontation against their strength.

Inducement: Encouraging someone to perform an act using rewards.

Infiltration: A maneuver to surreptitiously penetrate an enemy's security and defenses.

Initiative: A leader's freedom to impose their will on the opponent by creating conditions that allow them to strike at the time, place, and manner of their choosing, rather than reacting to circumstances shaped by the enemy.

Insurgent: Anyone taking part in a government overthrow who the established legal system does not recognize as a lawful combatant.

Intelligence: The act of collecting useful information, often regarding an opponent's forms of power.

Intelligence officer (also known as an operative): A member of an intelligence agency that recruits and manages agents.

Interdiction: The act of intercepting forces or supplies as they traverse a region. See **cordon**.

Intergalactic: Between different galaxies. See **interplanetary, interstellar**.

Interior lines: The maneuver corridors between forces enclosed within an area.

Interplanetary: Between planets in the same solar system. See **intergalactic, interstellar**.

Interstellar: Between different solar systems. See **intergalactic, interplanetary**.

Key terrain: Geographic locations and principles that provide one side with an advantage.

Kill zone: An area where a commander intends to strike enemy forces.

Kinetic weapon: A weapon that fires a non-explosive projectile, relying on the mass and velocity of the projectile to inflict damage.

Liaison officer: A representative from one group who is assigned to another group to communicate and coordinate between the two groups.

Line of communication (LOC): The route that connects a military unit with its supply base and enables the transit of troops, supplies, and information.

Local superiority: A temporary tactical advantage gained by concentrating an overwhelming force against an opponent at a decisive location to achieve a tactical advantage, despite broader inferiority.

Magic: An effect often credited to a person and not accepted or understood by science.

Main body: The primary component of a military force. See **ward**.

Maneuver: The movement of forces in relation to an enemy in a way that provides an advantage.

Martyr: Someone who sacrifices themselves or endures suffering for a cause or belief.

Means: The resources available to pursue ends.

Meeting engagement (also known as a chance encounter): A confrontation where military forces unexpectedly encounter each other.

Mercenary: Any private individual or group hired by another party to perform military operations for a fee.

Military power: A society's ability to use violence, or the threat of violence, to influence an opponent.

Militia: An army composed of citizens mobilized for an emergency.

Miracle: An event attributed to supernatural powers with no scientific explanation.

Mirror imaging: An analytical error of assuming an adversary will perceive situations, make decisions, and act according to one's own cultural norms, values, and reasoning rather than their own.

Mission creep: The gradual expansion of a military operation beyond its initial scope and objectives.

Mobilization: The steps taken to reallocate a society's resources for war.

Mothball: Deactivating, storing, and preserving equipment for potential future use.

Mop up: To complete an operation by killing or capturing the remaining enemy forces.

Neutrality: A political strategy where a party does not take a side in a conflict between other parties, and in response, hopes to avoid being attacked by either party. See **nonalignment**.

Non-alignment: A political strategy to distance oneself from any military alliances to preserve neutrality in case of war. See **neutrality**.

Non-combatant: Civilian and military personnel who do not participate in combat operations.

Objective: The object or idea whose destruction, neutralization, or capture enables attaining the desired ends.

Operation: The implementation of tactics in a region to accomplish military objectives that support the overall ends.

Order of battle: A listing or diagram depicting the strength and structure of a military force.

Overlapping fields of fire: The effect when two or more weapons can strike the same target.

Overt: Openly acknowledged activities.

Overthrow: A seizure of power.

Overwatch: A maneuver in which one unit moves while another provides security and protective fire.

Paper tiger: A person or organization that appears threatening but is ineffective or powerless.

Paramilitary: Armed groups that are not part of conventional armed forces of any group, but resemble them in organization, equipment, training, or mission.

Parasite: An organism that lives off a foreign host in a symbiotic or slave relationship.

Passive sensor: A detection device that uses the energy in the environment (ambient energy) to detect objects.

Peace treaty: A formal and permanent agreement to end hostilities.

Persuasion: Encouraging someone through reason to perform an act, often involving a benefit.

Perspective-taking: The process of understanding and appreciating the thoughts, feelings, motivations, and perspectives of others.

Perfidy: The deliberate act of betraying an enemy's trust by feigning protected status, such as pretending to be a civilian, medic, or surrendering combatant, to gain a tactical advantage.

Philosophy: A set of beliefs that alters and binds how a group thinks.

Picket: An individual or small group stationed outside a main body to provide security and early warning.

Poison pill: A defensive strategy to deter potential conquerors by presenting oneself as unattractive or undesirable.

Policy: Guidance that is directive, such as rules of engagement (ROE) that define the circumstances under which a military force may fire its weapons.

Political power: The ability to use official and unofficial means and ways to persuade, induce, or coerce others to act in ways that are beneficial to society and those who control the political establishment.

Power: The ability to obtain one's ends in the face of opposition.

Power sphere: The dynamic range of a society's influence, radiating from its **forms of power**, and shaping the behavior and alignment of neighbors.

Primitive army: A less organized and often irregular force, typically lacking advanced technology and formal training.

Private military company (PMC): See **mercenary.**

Privateer: Any individual granted a license by a government to conduct war in exchange for the spoils of war.

Prize court: A legal body that determines the value of captured assets and awards prize money.

Prize money: A reward given to warriors for capturing or securing an objective. See **prize court.**

GLOSSARY

Probe: Unmanned autonomous devices designed for exploration, data collection, and surveillance. See **droid.**

Professional army: A well-trained and organized military force composed of career soldiers with advanced weaponry and disciplined command structures.

Prominent individual: A person holding an influential position, such as a political or military leader.

Propaganda: The use of media to influence a group's beliefs, attitudes, or behaviors in support of an agenda.

Proportionality: Using the minimum necessary force to achieve an objective.

Psychological operations: The use of non-lethal means to influence the perceptions, emotions, and decision-making of individuals or groups to achieve strategic objectives.

Puppet government: A government that possesses the outward appearance of autonomy and authority but is under the control of an outside power.

Raid: A temporary incursion into enemy territory.

Reactionary: Anyone who opposes political, social, or cultural change in pursuit of a previous state. See **revolutionary.**

Rear-guard: A covering force positioned to protect the rear flank of a military element, especially during a retrograde or breakout. See **covering force, ward.**

Reconstitute: To restore a military unit to its authorized strength in personnel, equipment, or capability.

Religion: A system of behaviors, practices, world views, sacred texts, holy sites, ethical codes, and societal structures centered on a set of beliefs and principles about the universe, often involving the reverence for a higher or controlling power.

Reluctant encounter: A situation in which a force is driven into an unfavorable engagement through circumstance, miscalculation, or enemy action, such as being surrounded, flanked, or overrun.

Reparation: Compensation for real or imagined loss, damages, and injury.

Restraint: A rule or requirement that prohibits some action, such as limiting acts that might cause collateral damage.

Retirement: A planned and organized **retrograde** while not in contact.

Retrograde: A planned and organized movement away from an opponent.

Revolution in military affairs: A fundamental change in warfare often brought about by changes in technology.

Revolutionary: Anyone who promotes change. See **reactionary.**

Rout: A disorganized movement away from an opponent.

Salient: A bulge or penetration projecting into enemy territory.

Sapient: A being who possesses the cognitive abilities for rational thought. See **sentient.**

Scalpel operations: Military operations designed to eliminate individuals and groups with specific skills or knowledge.

Scorched Earth strategy: Deliberately eliminating all resources from a region to deny them to an opponent or their population.

Screen: A military force that provides surveillance and early warning to the main body.

Security: Operations that provide early and accurate warning of enemy and potential enemy operations.

Security dilemma: A situation that occurs when a society's actions intended to heighten its security lead others to respond with similar measures, producing increased tensions and conflict that neither side desires.

Sedition: Actions intended to incite rebellion or overthrow established authority. See **treason.**

Set-piece battle (also called a pitched battle): An operation where a leader deliberately places forces on chosen terrain to maximize their strengths against an opponent's weaknesses.

Sentient: A being that can access its senses. See **sapient.**

Shock and awe: A strategy that uses overwhelming power to demonstrate that continued resistance is futile.

Siege (also called a beleaguerment): The act of surrounding a position to compel its surrender or weaken it prior to an invasion through isolation and pressure. See **blockade.**

Sign: A message from a supernatural power.

Signature strikes: Military operations that target individuals or groups that exhibit specified behaviors.

Skirmish: A light force spread out to harass, delay, disrupt, or demoralize, using minor battles and avoiding direct conflict.

Society: A collection of individuals under a single government.

Spoiling attack: A limited attack to destroy, disable, or destabilize forces threatening an imminent attack.

Standoff weapon: A weapon that can strike a target from a distance while avoiding return fire.

Stratagem: A clever plan to outwit an opponent using subterfuge or unconventional ways.

Strategic obsolescence: The state in which a society or military becomes vulnerable due to its failure to adapt to evolving technologies, doctrines, or geopolitical realities.

Strategy: A society's long-term plan that links its ways and means to its desired ends.

Stronghold (also called a fortification): A military construction or position built to improve a defense.

Subversive: An individual or group that acts to overthrow an established power.

Supply chain: The system of means and ways involved in the creation and delivery of a product.

Suppressive fire (also known as covering fire): Weapon fire that degrades a target's effectiveness by threatening to strike them.

Surprise: A psychological condition whereby an opponent has insufficient time or resources to act effectively.

Surrender: Relinquishing control of military forces, populations, or terrain.

Tactics: The short-term decisions regarding the movement of forces and employment of weapons within a **battlespace** to implement a strategy.

Task force (also called strike group): A temporary grouping of military forces to conduct a specific operation or mission.

Technology: The combination of scientific knowledge, resources, and the means to fabricate useful artifacts.

Technological power: The ability of a society to apply scientific knowledge, resources, and manufacturing capabilities to create, maintain, and use complex artifacts, and thereby shape its environment, influence others, and assert dominance in political, economic, and military spheres.

Terrorism: The intentional use of violence or threats of violence against non-combatants to achieve political, ideological, or religious objectives.

Theocracy: A form of government that involves religious leaders or groups controlling a society's power.

Tit-for-tat: A strategy of cooperation until harmed, and then of retaliation in kind.

Total war: The commitment and mobilization of all aspects of society to a war effort.

Treason: Actions that aid an enemy or deliberately harm a society. See **sedition.**

Tribute: An act, statement, or gift intended to show subservience, gratitude, respect, or admiration.

Truce: An informal halt in hostilities, usually for a limited duration.

Turncoat: A person who shifts their allegiance from one group to another group.

Unconditional surrender: A capitulation whereby the victor unilaterally defines the terms. See **conditional surrender**.

Unrestricted warfare: Warfare that can use all aspects of power.

Vanguard: An advance or leading group. See **ward**.

War: The application of government-sponsored violence or the threat of violence in pursuing an end.

Ward: The components of a military force. The leading element is the vanguard (also called the advance guard or forward detachment), the middle is the main body, the flanks provide security on the periphery, and the rear-guard protects the trailing elements.

Warfare: A society's organized and systematic use of violence to force its will on another.

Warrior spirit: The personal attributes that make warriors effective, including courage in the face of fear, resiliency, fortitude in adversity, and an unwavering determination to overcome challenges.

Ways: The method that a society uses to apply means to achieve ends.

Weapon of mass destruction: Any weapon that creates devastating results; this may include nuclear weapons, biological weapons, and planet killers.

Withdrawal: A planned and organized **retrograde** while not in contact.

Wonder: Something that causes amazement and awe, often demonstrating the power of a religion.

Xenocide: The act of one species causing the extinction of another species.

Xenoforming: Deliberately converting an existing global ecosystem into one more suitable for a specific species. For humans, terraforming is deliberately modifying an inhospitable planet or moon to make it similar to Earth and more suitable for human habitation.

Xenophilia: The belief that your culture is superior to another culture.

Xenophobia: A real or imagined fear or hatred of any species different from one's own. See **ethnophobia**.

Xenovore: A creature that consumes an alien species.

ABOUT THE AUTHOR

Dr. Michael VanPutte (Lieutenant Colonel, U.S. Army, retired) is a strategist, technologist, and author who explores the future of conflict at the crossroads of military science, technology, and imagination. He served 24 years in the U.S. Army, beginning as an infantryman, completing Airborne and Ranger School, and later commanding combat engineers during Operation Desert Storm. After nearly a decade as a combat engineer officer, he transitioned to cyber operations, serving as Director of the U.S. Army AI Center and teaching strategy and technology at the U.S. Army War College, where he helped shape the next generation of senior military leaders. He then served as Deputy Director of Operations for a predecessor to U.S. Cyber Command and as a Program Manager at DARPA, leading advanced military research and development initiatives.

After retiring, Dr. VanPutte consulted for the Director of National Intelligence and the Department of Homeland Security, and co-founded Provatek LLC, a defense company specializing in advanced cyber operations. He has taught strategy at the U.S. Army War College and Worcester Polytechnic Institute, and computer security at The Ohio State University.

Dr. VanPutte earned a BS from The Ohio State University, an MS in Computer Science from the University of Missouri–Columbia, and a PhD in Computer Science from the Naval Postgraduate School. He writes both nonfiction and fiction, including works of military strategy, science fiction, and murder mysteries. His books and current projects are available at www.mvanputte.com.

www.ingramcontent.com/pod-product-compliance
Lightning Source LLC
Chambersburg PA
CBHW070309310726
48976CB00005B/1633